Self-energizing Electro Hydraulic Brake

Von der Fakultät für Maschinenwesen der Rheinisch-Westfälischen Technischen Hochschule Aachen zur Erlangung des akademischen Grades eines Doktors der Ingenieurwissenschaften genehmigte Dissertation

vorgelegt von

Matthias Liermann

aus

Essen

Berichter: Univ.-Prof. Dr.-Ing. Hubertus Murrenhoff
Univ.-Prof. Dr.-Ing. Dirk Abel

Tag der mündlichen Prüfung: 8. September 2008

Diese Dissertation ist auf den Internetseiten der Hochschulbibliothek online verfügbar.

Reihe Fluidtechnik

Band 49

Matthias Liermann

Self-energizing Electro-Hydraulic Brake

D 82 (Diss. RWTH Aachen University), 2008

Shaker Verlag
Aachen 2008

Bibliographic information published by the Deutsche Nationalbibliothek
The Deutsche Nationalbibliothek lists this publication in the Deutsche
Nationalbibliografie; detailed bibliographic data are available in the Internet at
http://dnb.d-nb.de.

Zugl.: Aachen, Techn. Hochsch., Diss., 2008

Copyright Shaker Verlag 2008
All rights reserved. No part of this publication may be reproduced, stored in a
retrieval system, or transmitted, in any form or by any means, electronic,
mechanical, photocopying, recording or otherwise, without the prior permission
of the publishers.

Printed in Germany.

ISBN 978-3-8322-7599-0
ISSN 1437-8434

Shaker Verlag GmbH • P.O. BOX 101818 • D-52018 Aachen
Phone: 0049/2407/9596-0 • Telefax: 0049/2407/9596-9
Internet: www.shaker.de • e-mail: info@shaker.de

Preface

This work developed during my activity as scientific staff at the Institute for Fluid Power Drives and Controls, RWTH Aachen University. I want to express my gratitude first of all to my doctoral advisor Professor Hubertus Murrenhoff for his support and mentorship. For the careful and critical review of this thesis I am grateful to Professor Dirk Abel, who accompanied this work within the context of the corporate research project "EABM". Thanks also to Professor Torsten Dellmann for his chairmanship in the doctoral examination procedure.

I want to thank Dr. Christian Stammen for his dedication in conducting the research of IFAS within the "EABM"-project. His scientific inspiration, encouragement and friendship were very valuable.

This work would have been impossible without the many supporters which accompanied it. In representation of all students, who participated as co-workers or doing their theses on the subject of this work I want to thank especially Julian Ewald, Titus Gotthardt, Bastian Scharre and Johannes Siegers for their creativity and commitment.

IFAS offered an inspiring atmosphere professionally and by its team spirit. I want to express my thanks to all, scientific, technical and administrative staff for this wonderful working atmosphere. Representatively I would like to name Marcell Meuser, Thomas Meindorf, Michael Zaun, Maxim Reichert, Torsten Verkoyen and Carlos Göhler who have been more than colleagues to me. Sharing our office together I especially enjoyed the company of Arshia Fatemi, his humor, his special view on things and the art of story telling.

Above all I thank God for his gift of faith, family and friends. My wife Barbara supports me in a wonderful way, our daughters Miriam and Simone are my source of joy.

Aachen, September 2008   Matthias Liermann

Selbstverstärkende Elektro-Hydraulische Bremse
Kurzfassung

Im Rahmen dieser Arbeit wurde ein neuartiges Bremskonzept, die Selbstverstärkende Elektro-Hydraulische Bremse (SEHB) erforscht. Die Bremse arbeit im Bereich der instabilen Selbstverstärkung, welche nur durch eine zusätzliche Regeleinrichtung technisch nutzbar wird. Das vorgestellte Schaltungskonzept ist auf dem Hintergrund einer Schienenfahrzeuganwendung entwickelt worden, ist aber grundsätzlich auf andere Anwendungen übertragbar. Vorteile des Konzepts sind unter anderem der minimierte Energieverbrauch, die Möglichkeit der Regelung des wirklichen Verzögerungsmoments und die Rückmeldefähigkeit über die ausschließlich elektrische Schnittstelle.

Die Arbeit stellt die SEHB herkömmlichen selbstverstärkenden Bremsen gegenüber und formuliert mit Hilfe statischer Betrachtungen die Grundlagen zur Charakterisierung der instabilen Selbstverstärkung. Zur Untersuchung der Dynamik der instabilen Selbstverstärkung wird ein linearisiertes Modell der ungeregelten Strecke entwickelt und auf Basis einer Poldominanzanalyse vereinfacht. Das vereinfachte Modell wird zur Bestimmung eines zustandsabhängigen Reglerkennfelds genutzt. Die grundlegenden hydraulischen Auslegungskriterien werden erläutert und eine Systematik der hydraulisch-mechanischen Ausführungsmöglichkeiten aufgestellt. Ein besonderer Fokus wird auf die Ventilansteuerung gerichtet, welche zentral für die Regelungseigenschaften der Bremse ist. Für eine erste Implementierung des Bremsenprinzips an Prototypen werden Ventile aus dem Kfz-Bereich (ABS, EHB, ASR und ESP) eingesetzt. Abschließend werden der Bremsenprüfstand und zwei Prototypen beschrieben, welche im Rahmen der Forschungsarbeit aufgebaut wurden. Verschiedene Ventilansteuerungsarten werden durch exemplarische Messergebnisse verglichen und diskutiert. Die erreichte Bremsdynamik der ersten Prototypen zeigt die Leistungsfähigkeit des neuen Bremsenprinzips auf.

Self-energizing Electro-Hydraulic Brake

Abstract

This thesis presents research results on a new fluid-mechatronic brake principle. The Self-energizing Electro-Hydraulic Brake (SEHB) utilizes the effect of instable self-reinforcement in combination with a closed loop control. Background for the development of the brake concept is a train application. However, SEHB is not limited to any specific application. Main advantages of the concept are its minimal energy consumption, the closed loop control of the true brake torque and its feedback ability due to the decentralized low-power electronic control.

This thesis introduces the new brake principle by comparing it to conventional self-reinforcing brakes. A mathematical distinction is given between self-reinforcement and self-energization on the basis of static considerations. The dynamic characteristics are analyzed using a linearized system description which is further simplified using the method of pole dominance analysis. The simplified model is used to calculate a state dependent proportional controller map on the basis of a damping criteria. Besides the theoretic analysis, the thesis presents the basic hydraulic design criteria and gives a systematic overview over different hydraulic-mechanical design solutions. A special focus is given on the valve control, since it is vital for the brake performance. Different automotive valves such as from anti-lock brake systems (ABS) or electronic stabilization programs (ESP) are applied using electronic power switches and current drivers. The brake test stand and two successive prototypes are outlined at the end of this thesis. Different exemplary measurement results show the performance of the implemented types of valve control and demonstrate the potential of this new brake technology.

Contents

Contents   I

Nomenclature   IV

1 Introduction   1
    1.1 Motivation: Why research on a new friction brake principle? 2
    1.2 Background of the research project on SEHB 3
    1.3 Structure of this thesis 6

2 Self-energizing brakes   8
    2.1 Principles of self-reinforcement and self-energization . 8
       2.1.1 Characteristic value C^* 13
    2.2 State of the art 13
       2.2.1 Drum brake 15
       2.2.2 Wedge Brake 18
             Electronic Wedge Brake (EWB) 19
    2.3 Self-reinforcing Hydraulic Brake 20
    2.4 Self-energizing Electro-Hydraulic Brake (SEHB) 24

3 Mathematical Model and Control Design — 28
3.1 Nonlinear model — 29
3.2 Linearization — 31
3.2.1 Valve — 31
3.2.2 Pressure dynamics of brake actuator — 33
3.2.3 Motion dynamics of brake actuator — 37
3.2.4 Pressure dynamics of supporting cylinder — 38
3.2.5 Motion dynamics of supporting cylinder — 42
3.2.6 Complete linearized model — 43
3.3 Analysis of pole configuration — 46
3.3.1 Comparing dominance of poles — 51
3.4 Simplified linear model — 54
3.4.1 Analysis of pole configuration — 59
3.5 Closed loop dynamics — 62
3.6 Adaptive proportional control — 67

4 Hydraulic-mechanical design of SEHB — 71
4.1 Design requirements — 72
4.1.1 General design requirements — 73
4.1.2 Application specific design requirements — 75
4.2 Hydraulics — 77
4.2.1 Cylinders — 78
4.2.2 Valves — 79
4.2.3 Accumulators — 80
4.3 Mechanics — 80
4.3.1 Brake actuator — 81
4.3.2 Actuator guidance — 83
4.3.3 Supporting cylinder type — 86
4.3.4 Supporting cylinder arrangement — 89
Alignment — 90
Mounting orientation — 91

Contents

5 Valve configurations for pressure control — 93
- 5.1 Valve arrangements — 93
- 5.2 Fast switching 2/2-valves — 95
  - 5.2.1 Pressure control using fast switching valves — 97
  - 5.2.2 Anti-lock brake system switching valves – characteristics and driver electronics — 99
- 5.3 Minimization of pressure steps — 103
  - 5.3.1 Throttle and bypass — 104
  - 5.3.2 Pulse modulation — 105
- 5.4 Proportional 2/2 valves — 109
  - 5.4.1 Measurement of tappet movement — 112
  - 5.4.2 Measurement of valve flow — 114
  - 5.4.3 Driver electronics — 116

6 Prototype design and tests — 121
- 6.1 Maximum brake force — 122
- 6.2 Brake test stand — 124
- 6.3 Prototype I, plunger actor — 125
  - 6.3.1 Hydraulic-mechanic design of Prototype I — 125
  - 6.3.2 Valve block of prototype I — 129
  - 6.3.3 Test results with prototype I — 131
    - Switching control — 132
    - Pulse frequency modulation — 134
    - Pulse width modulation — 136
- 6.4 Prototype II, differential actor — 138
  - 6.4.1 Hydraulic-mechanic design of Prototype II — 140
  - 6.4.2 Valve block of Prototype II — 142
  - 6.4.3 Test results with Prototype II — 143

7 Conclusion — 148

8 Bibliography — 151

Nomenclature

Symbol	Meaning	Unit
A	Area	m^2
A_{BA}	Piston area of brake actuator chamber A	m^2
a_{BA}	Acceleration of brake actuator piston	$\frac{m}{s^2}$
A_{Diff}	Differential Area of Piston	m^2
A_{sup}	Piston area of supporting cylinder	m^2
a_{sup}	Acceleration of supporting piston	$\frac{m}{s^2}$
A_{v}	Flow cross section of valve	m^2
α	Piston area ratio of differential cylinder	−
α_{D}	Flow coefficient of orifice	−
$C_{\text{BA}_{\text{A}}}$	Brake actuator capacity chamber A	$\frac{m^3 m^2}{N}$
$C_{\text{BA}_{\text{B}}}$	Brake actuator capacity chamber B	$\frac{m^3 m^2}{N}$
C_{brake}	Hydraulic capacity	$\frac{m^3 m^2}{N}$
c_{cal}	Caliper stiffness	$\frac{N}{m}$
C_{lp}	Capacity of low pressure line	$\frac{m^3 m^2}{N}$
$c_{\text{s}_{\text{sup}}}$	Stiffness of retraction springs in supporting cylinder	$\frac{N}{m}$
C^*	Brake coefficient $\frac{F_{\text{brake}}}{F_{\text{clamp}}}$	−
d	Diameter	m
d_{BA}	Viscous friction in brake actuator	$\frac{Ns}{m}$

Nomenclature

Symbol	Meaning	Unit
D_{cal}	Damping coefficient of caliper	$-$
d_{sup}	Viscous friction in supporting cylinder	$\frac{Ns}{m}$
Δp_{step}	Pressure step	$\frac{N}{m^2}$
F_{brake}	Brake force	N
F_{clamp}	External actuation force for brake mechanism	N
F_{d}	Deceleration force	N
$f_{e_{\text{cal}}}$	Mechanical natural frequency of caliper	$\frac{1}{s}$
$F_{f_{\text{BA}}}$	Friction force of brake actuator	N
$F_{f_{\text{sup}}}$	Friction force of supporting cylinder	N
F_{inc}	Additional force component added by self-reinforcement	N
F_{N}	Compression force of brake caliper	N
$F_{s_{\text{BA}}}$	Initialization spring force of brake actuator	N
$F_{s_{\text{sup}}}$	Retraction spring force of supporting cylinder	N
i_{L}	Transmission ratio $\frac{F_{\text{brake}}}{F_{\text{sup}}}$	$-$
I	Current	A
K_{cal}	Position factor of caliper	$\frac{m^3}{N}$
K_{sup_v}	Volume flow factor (dependency of v_{sup})	$\frac{m^3 \cdot s}{s \cdot m}$
K_{sup_y}	Volume flow factor (dependency of y)	$\frac{m^3}{s \cdot m}$
$K_{v_{\text{BA}}}$	Pressure build-up factor (dependency of v_{BA})	$\frac{N \cdot s}{m^2 s \cdot m}$
K_y	Pressure build-up factor (dependency of y)	$\frac{N}{m^2 s \cdot m}$
m_{BA}	Mass moved by load pressure in brake actuator	kg
m_{sup}	Mass of supporting piston and caliper	kg
μ	Friction coefficient	$-$
μ_{min}	Minimum friction coefficient	$-$
$\omega_{0_{\text{cal}}}$	Characteristic angular frequency	$\frac{N}{m^2}$
p_{BA_A}	Pressure in brake actuator chamber A	$\frac{N}{m^2}$
p_{BA_B}	Pressure in brake actuator chamber B	$\frac{N}{m^2}$
p_{BA_L}	Load pressure of brake actuator	$\frac{N}{m^2}$
p_{lp}	Pressure in low pressure line	$\frac{N}{m^2}$

Symbol	Meaning	Unit
p_{nom}	Nominal pressure used for valve characterization	$\frac{N}{m^2}$
p_{sup}	Pressure in pressurized chamber of supporting cylinder	$\frac{N}{m^2}$
p_{sup_L}	Load pressure of supporting cylinder	$\frac{N}{m^2}$
q	Ratio between additional force component F_{inc} and brake force F_{brake}	–
Q_{BA_A}	Volume flow into brake actuator chamber A	$\frac{m^3}{s}$
Q_{BA_B}	Volume flow from brake actuator chamber B	$\frac{m^3}{s}$
Q_{lp}	Flow into low pressure line	$\frac{m^3}{s}$
Q_{nom}	Nominal flow of valve at p_{nom}	$\frac{m^3}{s}$
Q_{sup}	Flow from the pressure side of supporting cylinder	$\frac{m^3}{s}$
R	Resistance	Ω
ρ	Densitiy	$\frac{kg}{m^3}$
T_A	Time between pulses	s
T_{BA_H}	Hydraulic time constant of brake pressure build-up	s
T_{BA_M}	Mechanic time constant of brake actuator	s
T_E	Pulse length	s
$T_{E,\text{dead}}$	Dead time of valve for switching on	s
$T_{O,\text{dead}}$	Dead time of valve for switching of	s
T_P	Cycle time	s
T_{sup_H}	Hydraulic time constant of supporting pressure build-up	s
T_{sup_M}	Mechanic time constant of supporting piston	s
u_0	Supply voltage	V
V_{acc}	Intake volume of high pressure accumulator	m^3
v_{BA}	Velocity of brake actuator	$\frac{m}{s^2}$
V_{BA_H}	Factor of pressure build-up in brake actuator	s

Nomenclature

Symbol	Meaning	Unit
$V_{\text{BA}_{\text{M}}}$	Velocity factor of actuator piston	$\frac{m}{N \cdot s}$
V_{hyd}	Gain of hydraulic path in simplified linear model of SEHB	–
V_{mech}	Gain of mechanical path in simplified linear model of SEHB	–
$V_{Q_{Y\text{BA},A}}$	Volume flow factor port A (dependency of y)	$\frac{m^3}{s}$
$V_{Q_{Y\text{BA},B}}$	Volume flow factor port B (dependency of y)	$\frac{m^3}{s}$
$V_{Q_{P\text{BA}_A}}$	Volume flow factor port A (dependency of p)	$\frac{m^3 m^2}{s \cdot N}$
$V_{Q_{P\text{BA}_B}}$	Volume flow factor port B (dependency of p)	$\frac{m^3 m^2}{s \cdot N}$
v_{sup}	Velocity of supporting piston	$\frac{m}{s^2}$
V_{sup_H}	Final value factor of supporting pressure	s
V_{sup_M}	Velocity factor of supporting piston	$\frac{m}{N \cdot s}$
x_{sup}	Position of supporting piston	m
x_{BA}	Position of brake actuator piston	m
y	Valve opening	100%

Chapter 1

Introduction

This thesis is about the development of a new brake principle, the self-energizing electro-hydraulic brake (SEHB). It is a friction brake which uses the brake force as the energy source to generate and control the compression force by electro-hydraulic force transmission. It has been invented at the Institute for fluid power drives and controls (IFAS, RWTH Aachen University) and developed within a research project funded by the German Research Foundation (DFG) since April of 2006. The scope of this thesis is to lay the foundations of mathematical modeling, control and mechanical-electrical design. First experimental results are presented which draw attention to the potential of this new brake principle.

1.1 Motivation: Why research on a new friction brake principle?

Friction brakes have some unique advantages besides the negative effect of wear. Friction is the most common principle used for braking. It transforms kinetic energy into heat due to the deformation and abrasion of the friction partner surfaces. A major advantage of friction brakes is that the friction force is only caused by the compression of the friction partners and not by their relative movement. Of course, movement and other state parameters have a significant influence on the friction coefficient. But a definite surface pressure always results in a definite deceleration down to stand still. No additional effect and no additional energy input is needed for maintaining the stand still, as long as the acceleration forces are smaller than the braking force. Friction brakes are very simple in their design and therefore especially favored in cases where security has high priority. A drawback, of course, is the wear produced during braking, which necessitates the replacement of friction material after a number of uses. The worn-off friction particles pollute the environment while the replacement causes machine down time and costs. For this reason wear-free brakes are advanced in an increasing number of applications. Nevertheless it is clear that, from the viewpoint of today, friction brakes will always be needed.

However, research is needed to make friction brakes fit for the future. Maintaining their high safety level, they need to be more efficient and more comfortable. The vehicle's kinetic energy, which is dissipated by the brakes, needs to be stored and reused in the best case. At least it should be used as energy source for the braking process. Using the brake as its own power plant avoids brake power distribution through the vehicle and reduces design complexity. There has been a trend from active safety systems over to automatic vehicle control systems over the past years which will certainly continue. These systems will clearly benefit from closed loop controlled brakes which give a feedback of the achieved deceleration irrespective of a changing friction coefficient.

1.2 Background of the research project on SEHB

The corporate research project in which the SEHB has been developed is funded by the German Research Foundation (Deutsche Forschungsgemeinschaft, DFG). The Institute of Automatic Control (IRT), the Institute of Rail Vehicles and Materials Handling Technology (IFS) and the Institute of Power Electronics and Electrical Devices (ISEA) are together with IFAS involved in this research project, which aims to develop a compact traction and braking module for an individual wheel on a rail vehicle. It has therefore been given the name "intelligent, integrated, independent wheel traction/braking module" or EABM (German: Einzelrad-Antriebs-Brems-Modul).

The distribution of traction and braking systems in a vehicle is such that their integration has posed a major challenge in the design of a rail vehicle up to now. The components required by the pneumatic braking systems include pressure reservoirs and switchboards, which frequently have to be fitted in the wagon because of the lack of space available in the undercarriage [Gra99]. The electronic traction power system is also separated from the motor, which is connected to the wheel set directly or via a transmission according to the traction concept. The number of interfaces to be taken into consideration include the lines that carry power to the motor, as well as the pneumatic supply lines for the brakes. This poses a problem as far as the design of the undercarriage is concerned, as these components are usually developed by different departments. The purpose of this project is therefore to develop an integrated traction and braking module that requires as few mechanical, data and electrical power interfaces as possible. **Fig. 1.1** illustrates the subject of the research project in the form of a sketch.

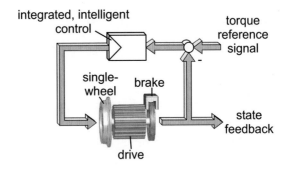

Figure 1.1: Intelligent, integrated, independent wheel traction/braking module (EABM)

The independent wheel module outlined in the lower part of the sketch is fitted in a closed-loop control circuit that realizes wheel slip and anti-skid braking, as well as lateral guidance [HHLS08]. Unlike the wheel sets that are usually used today, which comprise of two wheels securely joined together by a shaft, the developed module is intended to drive and brake just one wheel. This objective gives rise to requirements which cannot be met by a conventional pneumatic solution purely by virtue of the limited space available. Although hydraulic systems offer the significant advantage of higher force density [Kip95], this is offset by the disadvantage of a potentially environmentally harmful fluid medium. A central pressure supply to the undercarriage is out of the question for safety reasons and because of new problems that would arise with respect to interfaces to the undercarriage. IFAS has been therefore endeavoring to develop a completely new hydraulic brake concept that only requires an electric control interface to the outside with a low power level, whereby the energy needed to apply the brakes is generated by the braking process itself [SS06]. The principle of this brake was published for the first time in [LS06].

To illustrate the system simplification brought by independent self-energizing brakes compared to existing brake concepts, **Fig. 1.2** compares the setup of the modern electro-pneumatic train brake system (ep-brake) with the setup of a system using SEHBs. Modern train

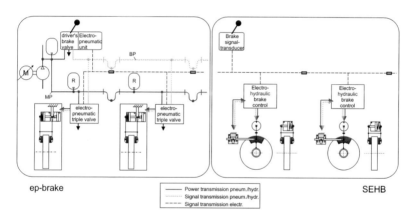

Figure 1.2: Power and signal transmission of electro-pneumatic brake compared to SEHB

brakes have one power and two signal lines through the whole train. The electrical line is supplemental to synchronize instantaneous braking of all brakes but it is not safety relevant. Safety is proved by the brake pipe (BP), which is directly connected to the driver's brake valve. If pressure drops in the brake pipe, the brakes are actuated.

The power for braking is supplied by the main reservoir pipe (MP) from a compressor usually installed in the locomotive. The brake power is accumulated locally in reservoirs in each train car. No power line is needed with brakes using the self-energizing effect. The self-energizing electro-hydraulic brake unit is supplied with low electric power. It is only needed for signal processing and valve control. The brake signal is transmitted electrically using a data bus in combination with a safety loop. It is locally controlled by each brake.

1.3 Structure of this thesis

This thesis presents the research results on SEHB over the past two years. It starts in chapter 2 with an explanation of how self-energization is connected to the principle of self-reinforcement, which has been used virtually since brakes have existed. The state of the art of brakes using a self-reinforcing principle is presented, going into more detail with the drum and the wedge brake. The working principle of SEHB is presented at the end of chapter 2.

The basic understanding of the brake's dynamics is based on a linearized system description developed in chapter 3. The linearized system equations are combined in the system transfer function and the state space representation. The state space representation is the basis for mathematical analysis of pole dominance. It appears that part of the system dynamics can be reduced. Therefore a reduced system dynamics representation is developed. This reduced system representation is used for controller development and analysis of closed loop stability.

Chapter 4 outlines the formulas used for designing the hydraulic components of the brake and presents the wide diversity of mechanical solutions, which can be realized regarding a specific application.

Introduction

A core part of the brake is the hydraulic control realized by a singular valve or a combination of valves. Chapter 5 presents and analyzes different solutions. Special focus is placed on combinations of seat valves, which mostly exist only as switching valves. After summarizing the literature on using switching seat valves in pressure control, the behavior and control of anti-lock brake system (ABS) valves used for SEHB prototype I are discussed. For Prototype II advanced valves of modern electronic stabilisation programs (ESP) are used which are presented at the end of chapter 5

Chapter 6 presents prototype designs and experimental results. Conclusions and outlook are given in Chapter 7

Chapter 2

Self-energizing brakes

Self-energizing brakes use only the vehicle's inertia to produce the desired brake force. The principle of self-energization goes further than the principle of self-reinforcement, because it not only amplifies an existing force but produces it out of itself. To clarify this, the chapter distinguishes both principles using the brake coefficient C^*, and gives typical examples of each principle. Finally the working principle of SEHB brake is introduced.

2.1 Principles of self-reinforcement and self-energization

Mechanical friction is the most fundamental way to dissipate kinetic energy from a moving mass. The classical approximation of the force of friction between two solid surfaces, F_{brake}, is known as Coulomb friction. It is proportional to the compressive force F_N by the friction coefficient μ which is just a simple parameter for characterization of a very complex cause-effect relationship [Per00].

$$F_\text{brake} = \mu F_\text{N} \tag{2.1}$$

The friction power P_f equals the friction force F_brake multiplied by the relative velocity v of the friction partners.

$$P_\text{f} = F_\text{brake} \cdot v \tag{2.2}$$

Therefore, to increase the friction power P_f, there are two options: increasing the relative velocity or the friction force. In some brake designs, through use of a transmission, a brake disk runs on brake shafts with a higher rotational speed than the actual drive shaft [BB04]. Other solutions aim at amplifying the actuation force via some kind of lever or hydro-static transmission. Often these solutions are combined with a brake booster that amplifies the input force using external energy sources. This type of increasing of the brake force is labeled as *servo-assistance*, in cases where external energy is added.

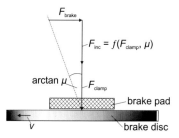

Figure 2.1: Self-reinforcement caused by feed-back of brake force

Self-reinforcing and self-energizing brakes make use of the friction force itself to intensify the compression. **Fig. 2.1** illustrates this. The braking force F_{brake} is partially the result of a feedback on itself. The clamping force F_{clamp} is just one force component of the total compression force $F_N = F_{\text{clamp}} + F_{\text{inc}}$.

To understand how the magnified brake force is built up, it is helpful to picture the process with a signal flow diagram. The process can be expressed feed-forward as an infinite series or as a feed-back loop, **Fig. 2.2**.

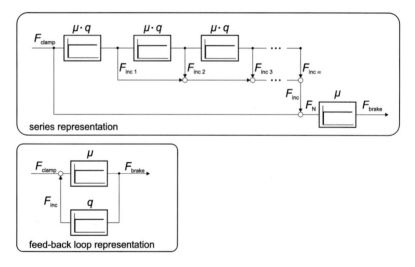

Figure 2.2: Flow diagram of self-reinforcement

The clamping force or input force F_{clamp}, which is produced hydraulically or mechanically, induces an initial braking force $F_{\text{clamp}} \cdot \mu$. Via some kind of transmission q, this adds a force increment F_{inc_1} on the compression. The added force component creates more friction

Self-energizing brakes

which is also transmitted to the compression by a factor q, resulting in another compression force increment F_{inc_2}. The sum of all F_{inc_n} results in F_{inc} as depicted in the upper diagram of Fig. 2.1, where the brake force is represented as the result of an infinite series. The factor q is:

$$q = \frac{F_{\text{inc}}}{F_{\text{brake}}} \qquad (2.3)$$

As the lower diagram of Fig. 2.1 shows, the reinforcement can be equally interpreted as an algebraic loop. The feedback gain of this loop can be derived mathematically from the series. The series of Fig. 2.2 can be written as

$$F_{\text{brake}} = [F_{\text{clamp}}(\mu q)^0 + F_{\text{clamp}}(\mu q)^1 + \ldots]\mu \qquad (2.4)$$

$$= \lim_{n \to \infty} \sum_{k=0}^{n} F_{\text{clamp}} \mu (\mu q)^n \qquad (2.5)$$

This series is well known as geometric series. The analytical solution is:

$$\sum_{k=0}^{n} F_{\text{clamp}} \mu (\mu q)^n = F_{\text{clamp}} \mu \frac{1 - (\mu q)^{n+1}}{1 - \mu q} \qquad (2.6)$$

For $n \to \infty$ and $\mu \cdot q < 1$ the solution converges to

$$F_{\text{brake}} = \lim_{n \to \infty} \sum_{k=0}^{n} F_{\text{clamp}} \mu (\mu q)^n = F_{\text{clamp}} \mu \frac{1}{1 - \mu q} \qquad (2.7)$$

In most brake designs, the compression force is acting on two sides of a brake disk or drum. This means that the brake force is doubled. It should be stressed here, that the reinforcement principle has been derived for single side friction contact only. For a double side friction contact, a factor of 2 is placed in Eq. 2.7.

From Eq. 2.7 one can see, that the clamping force can be magnified significantly, almost infinitely, by designing a brake with a $\mu \cdot q$ close to 1. However, μ is state dependent and can vary by a factor of 4. Therefore all conventional brakes are designed such that μq would not reach the value 1 even for very high friction coefficients.

For $\mu q > 1$ the brake force in Eq. 2.5 is not converging to a finite value. Even a very small input force F_{clamp} will lead to a theoretically infinite high brake force F_{brake}. In a vehicle the maximum brake force is limited by the peak traction between wheels and road or tracks. This unstable braking behavior has always been avoided by designers. Today however, drives and control devices have become more compact and are powerful enough to stabilize the process. The benefit of using this unstable brake behavior is that actuator forces can be much lower as in the case when the whole compression would have to be applied. The unstable brake is efficient because it is energizing itself.

It is helpful to distinguish between brake principles using the conventional self-reinforcement and the unstable self-energization. Therefore it is proposed to define self-reinforcing brakes as brakes with a stable feedback of the brake force on themselves ($\mu q < 1$). In contrast to this, self-energizing brakes shall be defined as brakes using the unstable feedback of the brake force on themselves ($\mu q > 1$). Brakes which work in both domains, dependent on the actual friction coefficient, are called semi-self-energizing brakes.

- **Self-reinforcing brakes**:
  Stable positive force feedback ($\mu q < 1$)

- **Self-energizing brakes**:
  Instable positive force feedback ($\mu q > 1$)

- **Semi-self-energizing brakes**: Self-reinforcing or self-energizing behavior dependent on friction coefficient ($\mu q < 1$, for low μ; $\mu q > 1$, for high μ)

2.1.1 Characteristic value C^*

The characteristic value C^*, sometimes referred to as brake shoe factor, generally defines the quotient of the effective peripheral or circumferential force F_{brake} and the input force F_{clamp}, generated hydraulically or mechanically [BB04].

$$C^* = \frac{F_{\text{brake}}(F_{\text{clamp}}, \mu)}{F_{\text{clamp}}} \qquad (2.8)$$

Inserting Eq. 2.5 and Eq. 2.6 yields:

$$C^* = \lim_{n \to \infty} (\mu \frac{1 - (\mu q)^n}{1 - \mu q}) \qquad (2.9)$$

With q only being dependent on the mechanical design. A high C^*-value leads to a huge compressive force generated by small input forces, which is generally desired. The higher the characteristic value, the greater also the sensitivity of the characteristic value on the friction coefficient

$$\frac{\partial C^*}{\partial \mu} = \lim_{n \to \infty} \frac{1 - (\mu q)^n [1 + n(\mu q - 1)]}{(\mu q - 1)^2} \qquad (2.10)$$

A small variation of the friction coefficient then has strong impact on the characteristic value which may lead to instability.

2.2 State of the art

This section gives an overview of the state of the art in self-reinforcing and semi-self-energizing brake principles. Many self-reinforcing brake principles have been developed in the history of brakes. The patent search conducted gives an overview about the various technical solutions for self-reinforcing brakes. A summary of this patent search is

Pat.No.	Year	Type	Actor	Effect of reinforcement	Controllable
DE 198 02 386 B4	2006	disc	mech.	pulled on sloped level combined with lever	self-regulating
DE 103 24 424 A1	2004	disc	electro-mech.	pad hinged on lever, friction force adds moment on lever	yes (by adapting angle of lever)
DE 102 01 555 A1	2003	disc	mecha-tronic	wedge	yes (closed loop)
DE 102 18 825 A1	2003	disc	mecha-tronic		yes (closed loop)
DE 195 39 012 A1	1997	drum	n.s.	pad hinged on lever, friction force adds moment on lever	yes (by adapting angle of lever)
DE 43 04 905 A1	1994	block	n. s.	Brake pad pulled along; a second pad is actuated via levers	no
DE 3725692 A1	1988	disc	mech.	rotation of brake pad on sloped level	no
DE 29 05 000 A1	1979	disc	hydr.	rotation of brake pad on sloped level	no
DE 1 775 941	1972	disc	mech.	brake pad pulled on sloped level	no
DE 15 30 869	1969	disc	hydr.	caliper pulled along; hydraulic force transmission	no
DE 1 228 152	1966	disc	hydr.	brake pad pulled on sloped level	no
DE 1 169 218	1964	disc	mech.	brake pad pulled on sloped level	no
DAS 1 131 531	1962	disc	hydr.	brake pad pulled on sloped level	no
DAS 1 112 017	1961	disc/block	mech.	brake pad pulled along; a block brake is actuated via levers	no
DAS 1 006 680	1957	disc	hydr.	pad hinged on lever, friction force adds moment on lever	no

Table 2.1: Patent search about self-reinforcing brakes

given by **Table 2.1**. It shows that actually very few different principles exist for self-reinforcing brakes. The basic idea is always the same: The actuation force is combined with the reinforcement via some kind of transmission, which is either made up of levers, a slider mechanism in combination with a sloped level, or a hydraulic force transmission. Patents of the past decade focus on solutions with self-regulating, adaptable or closed loop controlled reinforcement. They allow the use of smaller actuators, because the factor of reinforcement can be higher. The basic principles of reinforcement by levers (drum brake) and sloped level (wedge brake) are explained in detail in the following sections. The Self-energizing Electro-Hydraulic Brake, introduced in Section 2.4, is the first known true self-energizing brake with a stabilized behavior enforced by closed loop control.

2.2.1 Drum brake

Drum brakes consist of a rotating drum and curved-shaped brake shoes hinged on fixed points inside the drum. **Fig. 2.3** depicts two basic configurations, simplex (a) and duplex (b) drum brake according to [Rob94]. The brake shoes are expanded by hydraulic or mechanic actuation onto the drum-rubbing surface. The effect is that, depending on the direction of rotation of the drum, one (in case of simplex brake) or both (in case of duplex brake) of the brake shoes will be pulled along with the drum by the friction force, creating an additional torque on the shoe and therefore increasing the compression. **Fig. 2.4** depicts the acting forces on one brake shoe.

The tangent of the angle α defines the ratio between brake force F_{brake} and additional force component F_{inc} which is factor q, Eq. 2.3.

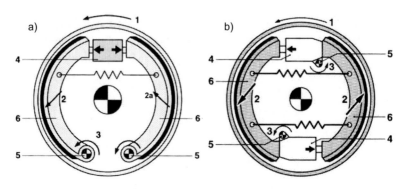

1. direction of motion 2. self-reinforcement of brake force 2a. self-reduction of brake force 3. torque 4. cylinder 5. pivoting point 6. brake shoe

Figure 2.3: Simplex (a) and duplex (b) drum brake according to [Rob94]

The brake coefficient C^* for one brake shoe follows directly from Eq. 2.9

$$C^*_{\text{drum}} = \lim_{n \to \infty} \mu \frac{1 - \frac{\mu}{\tan \alpha}^n}{1 - \frac{\mu}{\tan \alpha}} \tag{2.11}$$

Typically drum brakes are not self-energizing, so Eq. 2.11 converges and yields:

$$C^*_{\text{drum}} = \mu \frac{1}{1 - \frac{\mu}{\tan \alpha}} \tag{2.12}$$

For comparison, the characteristic value can also be derived directly from the forces marked in Fig. 2.4:

$$C^*_{\text{drum}} = \frac{F_{\text{brake}}}{F_{\text{clamp}}} = \frac{\mu(F_{\text{clamp}} + \frac{F_{\text{brake}}}{\tan \alpha})}{F_{\text{clamp}}} = \mu + \frac{\mu C^*_{\text{drum}}}{\tan \alpha} \tag{2.13}$$

$$\Leftrightarrow C^*_{\text{drum}} = \mu \frac{1}{1 - \frac{\mu}{\tan \alpha}} \tag{2.14}$$

Self-energizing brakes

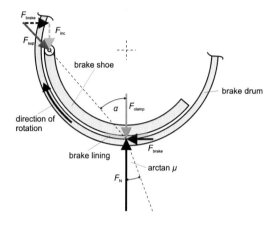

Figure 2.4: Principle of drum brake

The term in Eq. 2.14 has a pole at $\frac{\mu}{\tan \alpha} = 1$. For $\frac{\mu}{\tan \alpha} > 1$ it becomes negative, which is difficult to interpret physically. Strictly speaking, the characteristic value cannot be calculated for this case because the brake force does not converge, Eq. 2.5. The fact that Eq. 2.14 becomes negative is interpreted in some literature, as meaning that a negative clamping force is needed to reach a finite value for the brake force [RGH+04].

In case of two acting brake shoes, as in the duplex drum brake, the brake coefficient is doubled. The conventional drum brake design tries to avoid the area close to the pole. Designing the brake with a large enough angle α ensures $\frac{\mu}{\tan \alpha} < 1$ under all conditions [Ort04].

2.2.2 Wedge Brake

The principle of a wedge brake is shown in **Fig. 2.5**. The wedge is situated between the friction partner with (μ_f), for example a brake disc, and a wedge level with ($\mu_s \ll \mu_f$). The movement of the friction partners results in a friction force, which pulls the wedge further into the gap. For the following deliberations, the friction coefficient between wedge and slide μ_s shall ideally be zero.

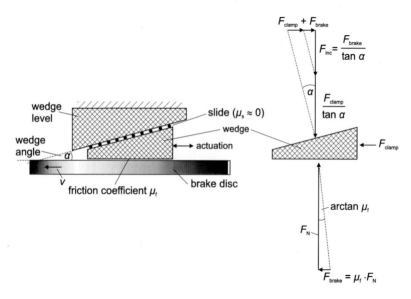

Figure 2.5: Principle of wedge brake

In equilibrium, the forces on the wedge are all in balance, as shown on the right side of Fig. 2.5. The horizontal actuation force F_{clamp} plus the friction force F_{brake} then equal the reaction force of the wedge level. Because of $\mu_s \approx 0$, the ratio between horizontal and vertical forces on the wedge is always $\frac{1}{\tan \alpha}$. The characteristic value of the wedge brake simply yields:

$$C^*_{\text{wedge}} = \frac{F_{\text{brake}}}{F_{\text{clamp}}} = \mu \frac{(F_{\text{clamp}} + F_{\text{brake}})}{\tan \alpha \cdot F_{\text{clamp}}} = \mu \frac{1 + C^*_{\text{wedge}}}{\tan \alpha} \quad (2.15)$$

$$\Leftrightarrow C^*_{\text{wedge}} = \frac{\mu}{\tan \alpha} \frac{1}{1 - \frac{\mu}{\tan \alpha}} \quad (2.16)$$

The wedge brake's characteristic value is very similar to that for the drum brake. As already explained above, Eq. 2.16 is only valid for $C^* > 0$.

It is also possible to derive the characteristic value directly from Eq. 2.9. The tangent of the angle α defines the ratio between brake force F_{brake} and additional force component F_{inc} which is the factor q, Eq. 2.3. In contrast to Fig. 2.2, the clamping force of the wedge brake is not acting directly on the compression but sideways onto the wedge. The resulting compression is the clamping force divided by $\tan \alpha$. Inserting in Eq. 2.9 yields:

$$C^*_{\text{wedge}} = \lim_{n \to \infty} \frac{\mu}{\tan \alpha} \frac{1 - \frac{\mu}{\tan \alpha}^n}{1 - \frac{\mu}{\tan \alpha}} \quad (2.17)$$

As said before, in most brake designs, the compression force is acting on two sides of a brake disk. In this case C^* is doubled.

Electronic Wedge Brake (EWB)

The wedge brake is one of the earliest self-reinforcing principles ever used. But it has gained some special attention since 1999, when the idea of a new electro-mechanical closed loop controlled wedge brake was first published [Sem99]. The idea was to reduce the energy consumption of an electric actuator by reducing the angle α, while a closed loop control stabilizes the self-reinforcing effect. The electronic wedge brake (EWB) is also known as eStop©. Its development is published in [HSPG01], [RSHG03], [RGH+04], [GG06], [FRBW+06]. For optimum performance, the EWB is designed to operate around the point at which the characteristic value just becomes infinite. This happens at an angle $\tan\alpha = \mu$. In this operating point the required control forces are minimal. The EWB is qualified as a semi-self-energizing brake because depending on the actual friction coefficient it must either pull or push the wedge.

At the ideal operating point, where the coefficient of friction is equal to the tangent of the wedge angle, the steady-state actuation force required to generate any braking torque is zero. From Eq. 2.16, it can be seen that for low coefficients of friction, C^* is positive, so a steady pushing force is required to maintain the braking force. When the coefficient of friction is greater than the tangent of the wedge angle, then a steady pulling force is required from the actuator to stop the wedge being pulled further in. **Fig. 2.6** shows the operating point of the EWB with an angle $\alpha = \arctan(\mu_{opt} = 0.352) = 19,4°$ in dependence of a variable friction coefficient. One can see that the actuator force at the optimal operating point is zero. Going to higher friction coefficients, a pulling force has to be applied to stabilize the brake. For a lower μ, the actor has to push the wedge into the friction contact.

2.3 Self-reinforcing Hydraulic Brake

All self-reinforcing principles have one thing in common. The tangential brake force is transmitted by some kind of transmission (in form of a lever, slide or gear) into a compressive force. Of course, hydrostatic transmission can also be used. The principle of a hydraulic

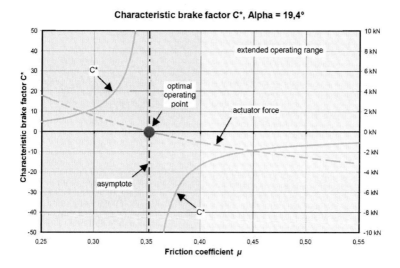

Figure 2.6: Characteristic brake factor C^* in relationship with the friction coefficient μ for a "push-pull-wedge" principle [HSPG01]

self-reinforcing brake is illustrated in **Fig. 2.7**. An external clamping force plus an additional pressure force, caused by pressure difference $p_{\text{sup}} - p_{\text{lp}}$ is summed up by a double rod brake actuator with piston area A_{BA} resulting in a compression force F_N. The additional pressure force equals F_{inc} in the above described brake principles. The pressure difference is caused by supporting the friction force F_{brake} via a supporting cylinder with piston area A_{sup}. Like the wedge brake concept, the compressing brake pad in the caliper must be movable in direction of the friction force. A hydraulic supporting piston connects the moving part to the bogie structure, fixing it between two

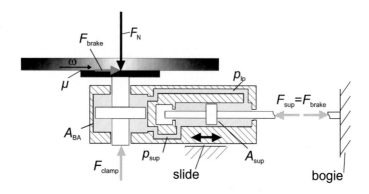

Figure 2.7: Principle of a self-reinforcing hydraulic brake

columns of oil. The ratio between added force component and brake force is the factor q, Eq. 2.3. Inserting q in Eq. 2.9 yields the characteristic value for a hydraulically self-reinforcing brake acting on one side of a brake disk:

$$C^*_{\text{hydr}} = \lim_{n \to \infty} \mu \frac{1 - \frac{\mu \cdot A_{\text{BA}}}{A_{\text{sup}}}^n}{1 - \frac{\mu \cdot A_{\text{BA}}}{A_{\text{sup}}}} \quad (2.18)$$

$$\Leftrightarrow C^*_{\text{hydr}} = \mu \frac{1}{1 - \frac{\mu \cdot A_{\text{BA}}}{A_{\text{sup}}}} \quad , \text{for} \quad \frac{\mu \cdot A_{\text{BA}}}{A_{\text{sup}}} < 1 \quad (2.19)$$

Usually the caliper encompasses the brake disk with two brake pads, so that the ratio between compression and brake force is doubled. Also, there may be a mechanical transmission that defines a ratio i_L between brake force and supporting force.

$$i_\text{L} = \frac{F_{\text{brake}}}{F_{\text{sup}}} \quad (2.20)$$

Fig. 2.8 depicts this set-up. It also shows a different configuration

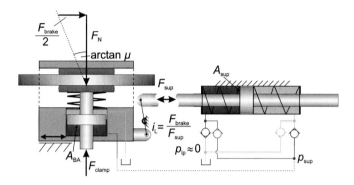

Figure 2.8: Self-reinforcing hydraulic brake with double sided actuation

with the supporting cylinder attached to the vehicle frame instead of the supporting piston as shown in the previous figure. The configuration of four check valves is acting as a hydraulic rectifier, which allows bi-directional service. The check valves assure that the lower line always conducts the high and the upper line the low pressure.

The ratio q between added pressure force component and brake force for the configuration in Fig. 2.8 yields:

$$q = \frac{F_{\text{inc}}}{F_{\text{brake}}} = \frac{p_{\text{sup}} A_{\text{BA}} \mu}{p_{\text{sup}} A_{\text{sup}} i_{\text{L}} \frac{1}{2}} = \frac{2 A_{\text{BA}} \mu}{i_{\text{L}} A_{\text{sup}}} \qquad (2.21)$$

Following from Eq. 2.9, the characteristic value of a hydraulic self-reinforcing brake considering double-sided actuation and mechanic transmission ratio i_{L} yields:

$$C^*_{\text{hydr}} = \lim_{n \to \infty} 2\mu \frac{1 - \frac{2\mu A_{\text{BA}}}{i_{\text{L}} A_{\text{sup}}}^n}{1 - \frac{2\mu A_{\text{BA}}}{i_{\text{L}} A_{\text{sup}}}} \qquad (2.22)$$

$$\Leftrightarrow C_{\text{hydr}}^* \;=\; 2\mu \frac{1}{1 - \frac{2\mu A_{\text{BA}}}{i_L A_{\text{sup}}}} \quad , \text{for} \quad \frac{2\mu A_{\text{BA}}}{i_L A_{\text{sup}}} < 1 \quad (2.23)$$

2.4 Self-energizing Electro-Hydraulic Brake (SEHB)

In searching for an idea for a new braking principle with minimized energy-consumption, it was considered to use the braking force as the energy source for hydraulic brake actuation. The pressure needed for actuation could be completely gained from hydraulic support of the friction force if the brake would be designed according to the self-energizing principle. A hydraulic control would be used in the hydraulic circuit of the self-reinforcing hydraulic brake (Fig. 2.8) to stabilize the self-energizing braking process via closed loop control. The idea of the <u>S</u>elf-energizing <u>E</u>lectro-<u>H</u>ydraulic <u>B</u>rake (SEHB) was born [SS06], which was first published in [LS06]. **Fig. 2.9** shows the working principle of SEHB.

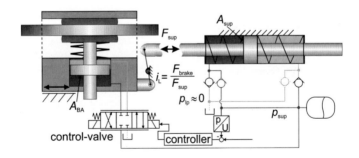

Figure 2.9: Principle of SEHB

In the SEHB concept there is no input or braking force anymore, which makes it difficult to define a characteristic value as in Eq. 2.8. Since it would be infinite anyway, the only possibility to characterize a self-energizing brake is by its dynamic behavior, which will be done in the following chapter. However, looking at the factor q, which determines the rate at which the brake force is self-energizing, reveals some interesting aspects:

$$q = \frac{F_{\text{inc}}}{F_{\text{brake}}} = \frac{F_{\text{SE}}}{F_{\text{brake}}} = \begin{cases} \frac{2\mu p_{\text{BA}} A_{\text{BA}}}{p_{\text{sup}} A_{\text{sup}} i_{\text{L}}} & \text{if valve opened positively} \\ 0 & \text{if valve closed} \\ \frac{-2\mu p_{\text{BA}} A_{\text{BA}}}{p_{\text{sup}} A_{\text{sup}} i_{\text{L}}} & \text{if valve opened negatively} \end{cases}$$
(2.24)

Self-energization and self-extinction The brake is self-energizing, when the control valve is positively opened. All forces remain constant when the valve is closed. In the opposite position of the control valve the process can be reversed. The supporting pressure then is used to actively pull back the brake actuator, which provides for a good dynamic performance for actuation and release of the brake. The following chapter will go into detail about the modeling and dynamic behavior of this process.

Control of actual retardation momentum Another interesting aspect following from Eq. 2.24 is that both pressures p_{sup} or alternatively p_{BA} can be used as closed loop control variables. While the necessity of a closed loop control might look like a drawback of SEHB at first, it is also one of the major advantages of SEHB: The supporting pressure is directly related to the friction force. With the supporting pressure used as control variable, the direct control of the actual braking force is possible, independently of friction coefficient changes. Conventional friction brakes only control the perpendicular actuator force. Since the friction coefficient μ is influenced by parameters like speed, brake pressure, temperature and environmental conditions like moisture or ice, conventional brakes can only roughly estimate the actual braking force F_brake.

The analysis in the following chapter will show that a proportional acting controller serves well to stabilize the hydraulic self-energization. Considering this fact, the feed-back loop can also be closed hydromechanically by comparison of the pressure in the supporting cylinder with a hydraulic or mechanic set value at the spool of the control valve. This idea can lead to a design without any electric components, [SLS06a].

Lifting of brake As common sense and Eq. 2.24 indicate, the process of self-energization stops when the supporting force becomes zero. Therefore, to actively pull the brake actuator back, some extra energy is needed. The amount of energy needed can be stored during the braking process in a hydraulic accumulator.

Engaging the brake when it is lifted In this case the supporting force is zero, $F_\text{sup} = 0$. For initiation of the self-energizing process, the contact between brake pad and disk has to be established by some mechanism. A simple way to do this without the need for an external energy source is to have a spring loaded when the brake pads are lifted from the brake disk. This spring then serves as accumulator to thrust the brake pad onto the disk when the hydraulic valve opens.

Retraction of supporting cylinder During each braking, the supporting cylinder moves a little bit. To bring it back into its middle position, it contains centering springs. By opening a connection between both chambers of the supporting cylinder, the supporting piston moves back to its initial position.

Tab. 2.2 summarizes some advantages and disadvantages of SEHB.

advantages	disadvantages
• no external hydraulic power supply	• more functional parts integrated in caliper
• closed loop brake control can deliver data about actual braking torque to superior control systems (autonomous braking)	• space required for moving parts of caliper
	• limited stroke of supporting cylinder
• constant braking torque irrespective of friction coefficient changes	

Table 2.2: Advantages and disadvantages of SEHB toward conventional hydraulic brakes

Chapter 3

Mathematical Model and Control Design

In this chapter the dynamics of self-energization is analyzed by methods for linear control systems. In the first part of the chapter the nonlinear system model is linearized. Then a simplified linearized model is obtained by applying the Litz method of pole dominance to identify the relevant dynamic subsystems. Pole configuration of the simplified model is interpreted to understand the process of hydraulic self-energization more deeply. It is also used for closed loop control design. In the last part of the chapter an example for an adaptive proportional control calculated on the basis of the simplified linearized model is validated by nonlinear system simulation.

3.1 Nonlinear model

The nonlinear hydraulic system is modeled according to the simplified scheme shown in **Fig. 3.1**.

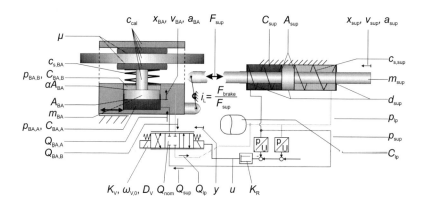

Figure 3.1: SEHB scheme for nonlinear system modeling

The corresponding signal flow diagram is shown in **Fig. 3.2**. It is divided by the shaded areas according to the system components valve, brake actuator and supporting cylinder. The mechanics of the cylinders and force transmission is modeled by Newton equations of motion. They include spring and damper characteristics. The spring characteristics are not only due to the springs depicted in the cylinders but predominantly by the stiffness of the caliper which acts as a spring. The damping is caused by (viscous) friction in the cylinders and bearings of the mechanism. The hydraulic capacity of the lines and cylinders is taken into account by pressure build-up equations. Flow inertia can be neglected because fluid acceleration

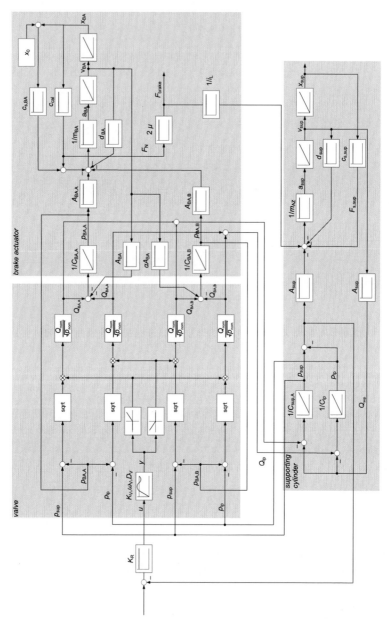

Figure 3.2: Signal flow diagram of nonlinear model of SEHB

is not influential for this application. The control is realized by a proportional feedback of the supporting pressure on the valve.

3.2 Linearization

The linearized system considers only system changes Δ around an operating point. A variable X can be expressed as the value X_0 at the operating point added by the change ΔX around the operating point.

$$X = X_0 + \Delta X \tag{3.1}$$

3.2.1 Valve

For positive opening, the valve depicted in Fig. 3.1 connects supporting pressure p_{sup} with piston face side A_{BA} of the brake actuator, Case A. The piston ring side αA_{BA} then is connected to low pressure line. Case B is the negative opening of the valve.

The direction of valve flow generally depends on the valve opening and the pressure difference $p_1 - p_2$ at its ports. The nonlinear flow equation of a sharp edged orifice is:

$$Q = A_{\text{v}} \sqrt{\frac{2}{\rho}} \alpha_D \sqrt{p_1 - p_2} \tag{3.2}$$

with A_{v} being the orifice flow cross section, ρ the specific weight of the fluid and α_D a flow parameter which typically is 0.6 for sharp edged orifices [Mur05]. Often a valve is characterized by the nominal flow Q_{nom} at a nominal pressure difference Q_{nom} over its ports when the valve is fully opened ($y = 100\%$). Assuming that the flow cross section is linear dependent on the valve opening, it is possible to write Eq. 3.2 as:

$$Q = \frac{Q_{\text{nom}}}{\sqrt{p_{\text{nom}}}} y \sqrt{p_1 - p_2} \qquad (3.3)$$

The linearization yields:

$$Q = Q_0 + \Delta Q(y, p_1, p_2) \qquad (3.4)$$

$$\Leftrightarrow Q = Q_0 + \left.\frac{\partial Q}{\partial y}\right|_{p_{1_0}, p_{2_0}} (y - y_0) +$$

$$+ \left.\frac{\partial Q}{\partial (\Delta p)}\right|_{y_0, p_{1_0}, p_{2_0}} (p_1 - p_{1_0} - p_2 + p_{2_0}) \qquad (3.5)$$

$$\Rightarrow \Delta Q = \left.\frac{\partial Q}{\partial y}\right|_{p_{1_0}, p_{2_0}} \Delta y +$$

$$+ \left.\frac{\partial Q}{\partial (\Delta p)}\right|_{y_0, p_{1_0}, p_{2_0}} (\Delta p_1 - \Delta p_2) \qquad (3.6)$$

$$\Leftrightarrow \Delta Q = V_{\text{Qy}} \Delta y + V_{\text{Qp}} (\Delta p_1 - \Delta p_2) \qquad (3.7)$$

with V_{Qy} being the flow factor expressing the dependence on valve opening and V_{Qp} for the dependence on change of pressure difference.

$$V_{\text{Qy}} = \frac{Q_{\text{nom}}}{\sqrt{p_{\text{nom}}}} \sqrt{p_{1_0} - p_{2_0}} \qquad (3.8)$$

$$V_{\text{Qp}} = \frac{Q_{\text{nom}}}{\sqrt{p_{\text{nom}}}} \frac{y_0}{2\sqrt{p_{1_0} - p_{2_0}}} \qquad (3.9)$$

The dynamics of the valve opening y is modeled by a second order lag system.

$$\frac{\partial \Delta y}{\partial^2 t} + 2 D_V \omega_{V_0} \frac{\partial \Delta y}{\partial t} + \omega_{V_0}^2 \Delta y = \omega_{V_0}^2 K_V \Delta u \qquad (3.10)$$

3.2.2 Pressure dynamics of brake actuator

The pressure build-up equations for actuator chambers A and B, neglecting external and internal leakage, are

$$\Delta \dot{p}_{BA_A} = \frac{1}{C_{BA_A}} [\Delta Q_{BA_A} - A_{BA} \Delta v_{BA}] \quad (3.11)$$

$$\Delta \dot{p}_{BA_B} = \frac{1}{C_{BA_B}} [-\Delta Q_{BA_B} + \alpha A_{BA} \Delta v_{BA}] \quad (3.12)$$

The flows $Q_{BA,A}$ and $Q_{BA,B}$ depend on the valve opening y and the pressure difference at the ports of the valve. The acting pressure difference depends on the opening direction of the valve.

Case A: Valve opened *positively* to increase braking pressure. From Eq. 3.7 follows the linearized flow for the process of increasing braking:

$$\begin{aligned}
\Delta Q_{BA_A} &= V_{Q y_{BA_A}} \Delta y + \\
&\quad + V_{Q p_{BA_A}} (\Delta p_{\text{sup}} - \Delta p_{BA_A}) \quad (3.13) \\
\Delta Q_{BA_B} &= V_{Q y_{BA_B}} \Delta y + \\
&\quad + V_{Q p_{BA_B}} (\Delta p_{BA_B} - \Delta p_{\text{lp}}) \quad (3.14)
\end{aligned}$$

Case B: Valve opened *negatively* to decrease braking pressure. The linearized flow equation for decreasing braking is:

$$\begin{aligned}
\Delta Q_{BA_A} &= V_{Q y_{BA_A}} \Delta y + \\
&\quad + V_{Q p_{BA_A}} (\Delta p_{BA_A} - \Delta p_{\text{lp}}) \quad (3.15) \\
\Delta Q_{BA_B} &= V_{Q y_{BA_B}} \Delta y + \\
&\quad + V_{Q p_{BA_B}} (\Delta p_{\text{sup}} - \Delta p_{BA_B}) \quad (3.16)
\end{aligned}$$

Case A will be studied in the following paragraphs. The solution for Case B is analogous and will be given afterwards.

The load pressure of the actuator is defined as:

$$\Delta p_{\text{BA}_\text{L}} = \Delta p_{\text{BA}_\text{A}} - \alpha \Delta p_{\text{BA}_\text{B}} \tag{3.17}$$

The change of actuator pressures $\Delta p_{\text{BA}_\text{A}}$ and $\Delta p_{\text{BA}_\text{B}}$ are expressed using the change of load pressure by solving the flow equations of a symmetric 4/3 way proportional valve [Fei87]:

$$\Delta p_{\text{BA}_\text{A}} = \frac{\Delta p_{\text{BA}_\text{L}}}{1+\alpha^3} \tag{3.18}$$

$$\Delta p_{\text{BA}_\text{B}} = -\frac{\alpha^2 \Delta p_{\text{BA}_\text{L}}}{1+\alpha^3} \tag{3.19}$$

Inserting Eq. 3.13 – 3.14 (Case A) and Eq. 3.17 – 3.19 into the pressure build-up equations Eq. 3.11 – 3.12 yields:

$$\Delta \dot{p}_{\text{BA}_\text{A}} = \frac{1}{C_{\text{BA}_\text{A}}} \Bigg[V_{\text{QyBA}_\text{A}} \Delta y + V_{\text{QpBA}_\text{A}} \Delta p_{\text{sup}} - \\ - V_{\text{QpBA}_\text{A}} \frac{\Delta p_{\text{BA}_\text{L}}}{1+\alpha^3} - A_{\text{BA}} \Delta v_{\text{BA}} \Bigg] \tag{3.20}$$

$$\Delta \dot{p}_{\text{BA}_\text{B}} = \frac{1}{C_{\text{BA}_\text{B}}} \Bigg[-V_{\text{QyBA}_\text{B}} \Delta y + V_{\text{QpBA}_\text{B}} \frac{\alpha^2 \Delta p_{\text{BA}_\text{L}}}{1+\alpha^3} + \\ + V_{\text{QpBA}_\text{B}} \Delta p_{\text{lp}} + \alpha A_{\text{BA}} \Delta v_{\text{BA}} \Bigg] \tag{3.21}$$

The load pressure build-up equation follows from Eq. 3.17:

$$\Delta \dot{p}_{\text{BA}_\text{L}} = \Delta \dot{p}_{\text{BA}_\text{A}} - \alpha \Delta \dot{p}_{\text{BA}_\text{B}} \tag{3.22}$$

Inserting Eq. 3.20 – 3.21 into Eq. 3.22 yields the differential equation for the load pressure build-up in the brake actuator:

$$T_{\text{BA}_\text{H}} \Delta \dot{p}_{\text{BA}_\text{L}} + \Delta p_{\text{BA}_\text{L}} = V_{\text{BA}_\text{H}} \bigg(K_\text{y} \Delta y - K_{\text{vBA}} \Delta v_{\text{BA}} +$$

Mathematical Model and Control Design

$$+ \frac{V_{\text{QPBA}_A}}{C_{\text{BA}_A}} \Delta p_{\text{sup}} -$$
$$- \alpha \frac{V_{\text{QPBA}_B}}{C_{\text{BA}_B}} \Delta p_{\text{lp}} \Bigg) \quad (3.23)$$

with the hydraulic time constant T_{BA_H} of the pressure build-up dynamics

$$T_{\text{BA}_H} = \left(\frac{V_{\text{QPBA}_A}}{C_{\text{BA}_A}} \frac{1}{1+\alpha^3} + \frac{V_{\text{QPBA}_B}}{C_{\text{BA}_B}} \frac{\alpha^3}{1+\alpha^3} \right)^{-1} \quad (3.24)$$

the final value V_{BA_H}

$$V_{\text{BA}_H} = T_{\text{BA}_H} \quad (3.25)$$

and the factors K_y and $K_{v_{\text{BA}}}$, which determine the influence of valve opening and pressure difference at the valve ports on the pressure build-up:

$$K_y = \frac{V_{\text{QyBA}_A}}{C_{\text{BA}_A}} + \frac{\alpha V_{\text{QyBA}_B}}{C_{\text{BA}_B}} \quad (3.26)$$

$$K_{v_{\text{BA}}} = \frac{A_{\text{BA}}}{C_{\text{BA}_A}} + \frac{\alpha^2 A_{\text{BA}}}{C_{\text{BA}_B}} \quad (3.27)$$

Additionally, the change of supporting pressure Δp_{sup} and the low pressure line pressure Δp_{lp} have influence on the dynamics. If the low pressure line is connected to a reservoir or low pressure accumulator, Δp_{lp} can be neglected.

For negative opening of the control valve (Case B), the differential equation of pressure built-up yields:

$$T_{\text{BA}_H} \Delta \dot{p}_{\text{BA}_L} + \Delta p_{\text{BA}_L} = V_{\text{BA}_H} \Bigg(K_y \Delta y - K_{v_{\text{BA}}} \Delta v_{\text{BA}} +$$
$$+ \alpha \frac{V_{\text{QPBA}_B}}{C_{\text{BA}_B}} \Delta p_{\text{sup}} -$$
$$- \frac{V_{\text{QPBA}_A}}{C_{\text{BA}_A}} \Delta p_{\text{lp}} \Bigg) \quad (3.28)$$

It is similar to Eq. 3.23. The difference is in the indices of flow factors and capacities in front of Δp_{sup} and Δp_{lp} and in the sign change of the volume flow factors V_{Qp} because $y_0 < 0$. This also affects T_{BA_H} and V_{BA_H}. Factors K_y and K_{vBA} are the same. The hydraulic time constant T_{BA_H} and the final value V_{BA_H} are:

$$T_{\text{BA}_\text{H}} = \left(-\frac{V_{\text{QPBA}_\text{A}}}{C_{\text{BA}_\text{A}}}\frac{1}{1+\alpha^3} - \frac{V_{\text{QPBA}_\text{B}}}{C_{\text{BA}_\text{B}}}\frac{\alpha^3}{1+\alpha^3}\right)^{-1} \quad (3.29)$$

$$V_{\text{BA}_\text{H}} = T_{\text{BA}_\text{H}} \quad (3.30)$$

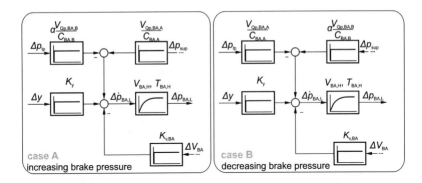

Figure 3.3: Linearized brake actuator pressure dynamics

The differential equations Eq. 3.23 and Eq. 3.28 have first order lag dynamics. **Fig. 3.3** depicts the signal flow diagrams for increasing and decreasing brake pressure.

3.2.3 Motion dynamics of brake actuator

The movement of the brake actuator piston is described by Newton's equation of motion.

$$m_{BA}\Delta a_{BA} = A_{BA}\Delta p_{BA_L} - d_{BA}\Delta v_{BA} - c_{cal}\Delta x_{BA} \quad (3.31)$$

$$\Leftrightarrow T_{BA_M}\Delta a_{BA} + \Delta v_{BA} = V_{BA_M}(A_{BA}\Delta p_{BA_L} - c_{cal}\Delta x_{BA}) \quad (3.32)$$

with the mechanic time constant of the brake actuator piston T_{BA_M} and the velocity factor V_{BA_M}:

$$T_{BA_M} = \frac{m_{BA}}{d_{BA}} \quad (3.33)$$

$$V_{BA_M} = \frac{1}{d_{BA}} \quad (3.34)$$

Eq. 3.32 can also be written in the characteristic way for second order lag systems:

$$\frac{A_{BA}}{m_{BA}}\Delta p_{BA_L} = \Delta a_{BA} + \frac{d_{BA}}{m_{BA}}\Delta v_{BA} +$$
$$+ \frac{c_{cal}}{m_{BA}}\Delta x_{BA} \quad (3.35)$$

$$\Leftrightarrow \omega_{0_{cal}}^2 K_{cal}\Delta p_{BA_L} = \Delta a_{BA} + 2D_{cal}\omega_{0_{cal}}\Delta v_{BA} +$$
$$+ \omega_{0_{cal}}^2 \Delta x_{BA} \quad (3.36)$$

The characteristic angular frequency $\omega_{0_{cal}}$, damping coefficient D_{cal} and the factor for the final position K_{cal} of the caliper are:

$$\omega_{0_{cal}} = \sqrt{\frac{c_{cal}}{m_{BA}}} \quad (3.37)$$

$$D_{cal} = \frac{d_{BA}}{2\sqrt{m_{BA}c_{cal}}} \quad (3.38)$$

$$K_{cal} = \frac{A_{BA}}{c_{cal}} \quad (3.39)$$

The natural frequency $f_{e_{cal}}$, often referred to as eigenfrequency, is where the system is most sensitive to excitation. It is the frequency of maximum amplitude in the bode diagram. For a system with damping, the natural frequency is always lower than the characteristic frequency.

$$f_{e_{cal}} = \frac{1}{2\pi}\omega_{0_{cal}}\sqrt{1 - D_{cal}^2} \qquad (3.40)$$

$$\Leftrightarrow = \frac{c_{cal}}{2\pi\, m_{BA}}\sqrt{\frac{4m_{BA} - d_{BA}^2}{m_{BA}}} \qquad (3.41)$$

Fig. 3.4 depicts the signal flow diagram of the dynamic behavior of the mechanical subsystem of the caliper. It is separated into a first oder lag system and an integrator according to Eq. 3.32.

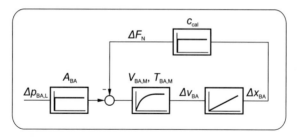

Figure 3.4: Dynamics of mechanical compression of caliper

3.2.4 Pressure dynamics of supporting cylinder

The pressure build-up equations for supporting chamber and low pressure side, neglecting external and internal leakage, are

$$\Delta\dot{p}_{sup} = \frac{1}{C_{sup}}[-\Delta Q_{sup} + A_{sup}\Delta v_{sup}] \qquad (3.42)$$

Mathematical Model and Control Design

$$\Delta \dot{p}_{\text{lp}} = \frac{1}{C_{\text{lp}}} [\Delta Q_{\text{lp}} - A_{\text{sup}} \Delta v_{\text{sup}}] \qquad (3.43)$$

The pressure dynamics in the low pressure line can be neglected with the use of a reservoir because $C_{\text{lp}} \gg C_{\text{sup}}$. The flow Q_{sup} depends on the valve opening y and the pressure difference at the ports of the valve. The acting pressure difference depends on the opening direction of the valve. The linearization of the valve flow equation for increasing braking yields:

Case A: Valve opened *positively* to increase braking pressure.

$$\Delta Q_{\text{sup}} = V_{\text{QYBA}_\text{A}} \Delta y + \\ + V_{\text{QPBA}_\text{A}} (\Delta p_{\text{sup}} - \Delta p_{\text{BA}_\text{A}}) \qquad (3.44)$$

Case B: Valve opened *negatively* to decrease braking pressure.

$$\Delta Q_{\text{sup}} = -V_{\text{QYBA}_\text{B}} \Delta y - \\ - V_{\text{QPBA}_\text{B}} (\Delta p_{\text{sup}} - \Delta p_{\text{BA}_\text{B}}) \qquad (3.45)$$

The signs must be inverted compared to Eq. 3.44 because the flow direction is always positive as indicated in Fig. 3.1. The supporting cylinder is the pressure source for the brake.

The following paragraphs derive the pressure equation for Case A. The solution for Case B is similar and is given at the end of the section.

The change of load pressure of the supporting cylinder is equal to the change of supporting pressure because of the large capacity of the low pressure line:

$$\Delta p_{\text{sup}_\text{L}} = \Delta p_{\text{sup}} - \Delta p_{\text{lp}} \qquad (3.46)$$

$$\Leftrightarrow \Delta p_{\text{sup}_\text{L}} = \Delta p_{\text{sup}} \quad \text{with } \Delta p_{\text{lp}} \stackrel{!}{=} 0 \qquad (3.47)$$

$$\Leftrightarrow \Delta \dot{p}_{\text{sup}_\text{L}} = \Delta \dot{p}_{\text{sup}} \qquad (3.48)$$

Inserting Eq. 3.44 and Eq. 3.42 into Eq. 3.48 yields the differential equation for the change of load pressure in the supporting cylinder:

$$\Delta \dot{p}_{\text{supL}} = \frac{1}{C_{\text{sup}}} \left[-V_{\text{QYBA}_A} \Delta y - V_{\text{QPBA}_A} p_{\text{sup}} + \right.$$
$$\left. + V_{\text{QPBA}_A} \Delta p_{\text{BA}_A} + A_{\text{sup}} \Delta v_{\text{sup}} \right] \quad (3.49)$$

Rearrangement yields:

$$T_{\text{supH}} \Delta \dot{p}_{\text{supL}} + p_{\text{supL}} = V_{\text{supH}} \left(-K_{\text{sup}_y} \Delta y + K_{\text{sup}_v} \Delta v_{\text{sup}} + \right.$$
$$\left. + \frac{V_{\text{QPBA}_A}}{C_{\text{sup}}} \Delta p_{\text{BA}_A} \right) \quad (3.50)$$

with the hydraulic time constant T_{supH} of the pressure build-up dynamics

$$T_{\text{supH}} = \frac{C_{\text{sup}}}{V_{\text{QPBA}_A}} \quad (3.51)$$

the final value V_{supH}

$$V_{\text{supH}} = T_{\text{supH}} \quad (3.52)$$

and the factors K_{sup_y} and K_{sup_v}, which determine the influence of valve opening and pressure difference at the valve ports on the pressure build-up:

$$K_{\text{sup}_y} = \frac{V_{\text{QYBA}_A}}{C_{\text{sup}}} \quad (3.53)$$

$$K_{\text{sup}_v} = \frac{A_{\text{sup}}}{C_{\text{sup}}} \quad (3.54)$$

For negative opening of the control valve (Case B), the differential equation of pressure built-up yields:

$$T_{\text{supH}} \Delta \dot{p}_{\text{supL}} + p_{\text{supL}} = V_{\text{supH}} \left(-K_{\text{sup}_y} \Delta y + K_{\text{sup}_v} \Delta v_{\text{sup}} + \right.$$

Mathematical Model and Control Design page 41

$$+ \frac{V_{\mathrm{QPBA_B}}}{C_{\mathrm{sup}}}(-\Delta p_{\mathrm{BA_B}})\bigg) \quad (3.55)$$

The main difference to Eq. 3.50 is that the sign in front of $\Delta p_{\mathrm{BA_B}}$ is changed. Factor $K_{\mathrm{sup_v}}$ remains the same. Valve opening factor $K_{\mathrm{sup_y}}$, hydraulic time constant T_{supH} and final value V_{supH} are:

$$K_{\mathrm{sup_y}} = -\frac{V_{\mathrm{QyBA_B}}}{C_{\mathrm{sup}}} \quad (3.56)$$

$$T_{\mathrm{supH}} = -\frac{C_{\mathrm{sup}}}{V_{\mathrm{QPBA_B}}} \quad (3.57)$$

$$V_{\mathrm{supH}} = T_{\mathrm{supH}} \quad (3.58)$$

The differential equations Eq. 3.50 and Eq. 3.55 have first order lag dynamics. **Fig. 3.5** depicts the signal flow diagrams of the supporting pressure for increasing and decreasing pressure.

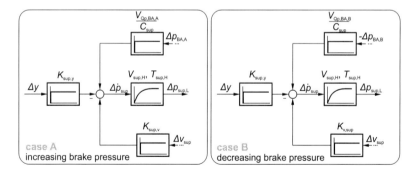

Figure 3.5: Linearized supporting cylinder pressure dynamics

3.2.5 Motion dynamics of supporting cylinder

The movement of the supporting piston and the attached caliper is described by Newton's equation of motion.

$$m_{\text{sup}}\Delta a_{\text{sup}} = -A_{\text{sup}}\Delta p_{\text{sup}_L} - d_{\text{sup}}\Delta v_{\text{sup}}$$
$$-c_{\text{s}_{\text{sup}}}\Delta x_{\text{sup}} + \frac{2\mu}{i_L}F_N \quad (3.59)$$

$$\Leftrightarrow T_{\text{sup}_M}\Delta a_{\text{sup}} + \Delta v_{\text{sup}} = V_{\text{sup}_M}(-A_{\text{sup}}\Delta p_{\text{sup}_L}$$
$$-c_{\text{s}_{\text{sup}}}\Delta x_{\text{sup}} + \frac{2\mu}{i_L}F_N) \quad (3.60)$$

with the mechanic time constant of the moved piston including attached caliper T_{sup_M} and the velocity factor V_{sup_M}:

$$T_{\text{sup}_M} = \frac{m_{\text{sup}}}{d_{\text{sup}}} \quad (3.61)$$

$$V_{\text{sup}_M} = \frac{1}{d_{\text{sup}}} \quad (3.62)$$

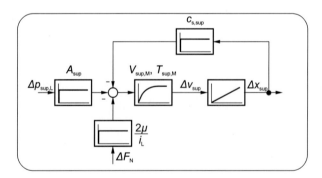

Figure 3.6: Supporting cylinder movement dynamics

Fig. 3.6 shows the signal flow diagram of the dynamic behavior of the supporting cylinder movement.

3.2.6 Complete linearized model

The complete linearized signal flow diagram of SEHB is shown in **Fig. 3.7** and **Fig. 3.8**.

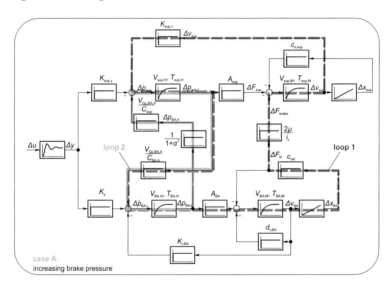

Figure 3.7: Complete signal flow diagram for increasing brake force

One can see the interlaced subsystems of brake actuator and supporting cylinder hydraulics and mechanics. The differences of both diagrams are marked in Fig. 3.8. One of the feedback loops (loop 1) runs from the brake actuator pressure dynamics through the Newton

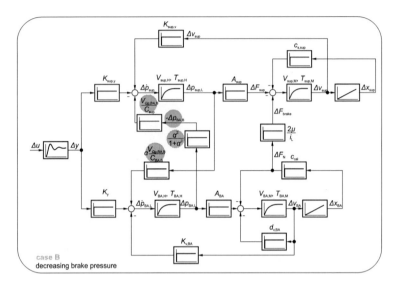

Figure 3.8: Complete signal flow diagram for decreasing brake force

equations of the caliper, where pressure is transformed into a brake force. The brake force is summed up in the motion equation of the supporting cylinder and the resulting flow creates a pressure rise in the supporting cylinder. The load pressure of the supporting cylinder multiplied with the flow factor influences the brake pressure. Since the volume flow factor $V_{Q_{PBA_B}}$ is positive for a positively opened valve (Eq. 3.9), the pole of loop 1 is positive and therefore instable. For the negatively opened valve, however, it is stable, because $V_{Q_{PBA_B}}$ is negative.

Another loop (loop 2) runs in parallel from the actuator load pressure to the pressure build-up of the supporting pressure. The supporting pressure, in turn, is fed back on the pressure build-up of the brake actuator. The numeric solution in the next section shows, that for positive opening of the valve, both poles are in the right half-plane and for negative valve opening both are in the left half plane of the pole diagram. The simplified linearized model in section 3.4 makes it easier to see the influences of both loops.

The transfer function of the linearized system is obtained by applying the Laplace transformation on the differential equations Eqs. 3.10, 3.23, 3.28, 3.32, 3.50, 3.55 and 3.60. Then the equations can be combined to eliminate all variables except for the variables of input $\mathcal{L}(\Delta u)$ and output $\mathcal{L}(\Delta p_{\mathrm{supL}})$. The initial values of the Laplace transformed variables can be set to zero. The transfer function of the complete system is of the form:

$$\frac{\mathcal{L}(\Delta p_{\mathrm{supL}})}{\mathcal{L}(\Delta u)} = \frac{a_5 s^5 + a_4 s^4 + a_3 s^3 + a_2 s^2 + a_1 s + a_0}{(b_8 s^2 + b_7 s)(b_6 s^6 + b_5 s^5 + b_4 s^4 + b_3 s^3 + b_2 s^2 + b_1 s + b_0)} \quad (3.63)$$

The polynomial order n of the denominator is 8. The first term in the denominator contains the valve dynamics with factors b_8 and b_7. It is separable from the rest because it is connected in series with the rest of the model. The polynomial order of the numerator is 5, so the difference order of the SEHB without valve is 1.

The state space representation of the linear system without valve dynamics is:

$$\dot{\mathbf{x}} = \mathbf{A} \cdot \mathbf{x} + \mathbf{b} \cdot \Delta y \quad (3.64)$$

$$\Delta p_{\mathrm{supL}} = \mathbf{c} \cdot \mathbf{x} \quad (3.65)$$

with

$$\mathbf{x} = \begin{bmatrix} \Delta p_{\text{BAL}} \\ \Delta x_{\text{BA}} \\ \Delta v_{\text{BA}} \\ \Delta p_{\text{supL}} \\ \Delta x_{\text{sup}} \\ \Delta v_{\text{sup}} \end{bmatrix} ; \mathbf{A} = \begin{bmatrix} a_{1,1} & 0 & a_{1,3} & a_{1,4} & 0 & 0 \\ 0 & 0 & a_{2,3} & 0 & 0 & 0 \\ a_{3,1} & a_{3,2} & a_{3,3} & 0 & 0 & 0 \\ a_{4,1} & 0 & 0 & a_{4,4} & 0 & a_{4,6} \\ 0 & 0 & 0 & 0 & 0 & a_{5,6} \\ 0 & a_{6,2} & 0 & a_{6,4} & a_{6,5} & a_{6,6} \end{bmatrix}$$

$$\mathbf{b} = \begin{bmatrix} b_1 \\ 0 \\ 0 \\ b_4 \\ 0 \\ 0 \end{bmatrix} ; \mathbf{c} = \begin{bmatrix} 0 & 0 & 0 & 1 & 0 & 0 \end{bmatrix} \tag{3.66}$$

3.3 Analysis of pole configuration

The goal of this section is to determine the dominating dynamics of the system. This allows further simplification of the model. Despite the fact that for each subsystem time constants have been defined, they cannot be compared with each other at this point. The influence of the subsystems on each other must also be considered. For example a change of the actuator position of only $0.001\ m$ results in a change of the supporting pressure of $6.7\ MPa$.

Poles are the roots of the denominator of the transfer function. They are also the eigenvalues of the system matrix $\mathbf{A}$. The pole configuration of linear systems provides insight about dynamic characteristics, namely the stability [Abe06]. Because of the principle of superposition, the step response $h(t)$ of a linear time invariant system is the sum of the step responses caused by each pole ν_k of the system. If the system is of order n and has only single poles $\neq 0$ it can be written:

$$h(t) = p_0 + \sum_{k=1}^{n} p_k e^{\nu_k t} \quad (3.67)$$

The factor p_k is constant if the eigenvalues are all single, as it is often the case in real systems. If one of the poles ν_k is positive, it is obvious that the step response becomes an unbounded function of time. Often, as is the case for any spring-damper system, ν_k is complex conjugate. This correlates to the oscillatory behavior of these systems, since

$$cos(t) = \frac{e^{jt} + e^{-jt}}{2} \quad (3.68)$$

However, if the real part of the conjugate complex pole pair is positive, the amplitude of the oscillation will increase exponential with time. For $\Re(\nu) = 0$ and $\Im(\nu) = 0$ the system is integrating, for $\Re(\nu) = 0$ and $\Im(\nu) \neq 0$ the oscillation is stabilized without damping and will not decease. **Fig. 3.9** illustrates the behavior of different pole configurations. For each pole the graph shows the typical dynamic behavior.

To get the pole configuration of the linearized SEHB model, a polynomial of the eighth order has to be solved. This is not possible analytically. Inserting values for the parameters and the operating points of the linearized flow factors a numerical solution can be computed though. **Table 3.1** lists values for parameters which stem from the brake design of the SEHB prototype II.

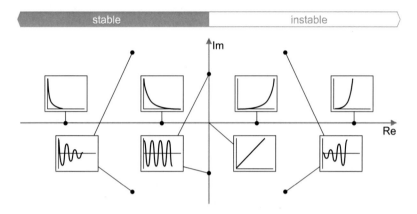

Figure 3.9: Effect of different pole configurations on system behavior

Expression	Value	Unit	Expression	Value	Unit
A_{BA}	$5.027 \cdot 10^{-3}$	m^2	d_{BA}	1.0	$\frac{Ns}{m}$
A_{sup}	$7.658 \cdot 10^{-4}$	m^2	d_{sup}	1.0	$\frac{Ns}{m}$
α	0.609		D_V	0.7	
C_{BA_A}	$7.181 \cdot 10^{-14}$	$\frac{m^5}{N}$	K_V	1.0	
C_{BA_B}	$4.373 \cdot 10^{-14}$	$\frac{m^5}{N}$	i_L	2.045	
c_{cal}	$1.5 \cdot 10^7$	$\frac{N}{m}$	m_{BA}	10.0	kg
C_{sup}	$1.644 \cdot 10^{-13}$	$\frac{m^5}{N}$	m_{sup}	12.0	kg
$c_{s_{BA}}$	$1.6 \cdot 10^4$	$\frac{N}{m}$	μ	0.35	
$c_{s_{sup}}$	$2.0 \cdot 10^3$	$\frac{N}{m}$	ω_{V_0}	$6.283 \cdot 10^2$	$\frac{Rad}{s}$

Table 3.1: Parameter values for SEHB

The operating point of linearization depends on the actual brake force and the valve opening, Eq. 3.7. All system pressures can be expressed as a function of the supporting pressure, [Fei87].

$$p_{sup} = \frac{F_{brake}}{A_{sup} i_L} \qquad (3.69)$$

$$p_{\text{BA}_L} = \frac{F_{\text{brake}}}{A_{\text{BA}} 2\mu} \tag{3.70}$$

$$p_{\text{BA}_B} = \frac{p_{\text{sup}} - \alpha^2 p_{\text{BA}_L}}{1 + \alpha^3} \tag{3.71}$$

$$p_{\text{BA}_A} = \frac{\alpha p_{\text{sup}} + p_{\text{BA}_L}}{1 + \alpha^3} \tag{3.72}$$

Using these relations, the transfer function can be expressed in dependence on the brake force and the valve opening. **Table 3.2** shows the system pressures for three different brake forces. Inserting these

F_{brake} [kN]	p_{sup} [bar]	p_{BA_A} [bar]	p_{BA_B} [bar]
2	12.77	8.70	10.98
4	25.54	17.40	21.96
6	38.31	26.09	32.94

Table 3.2: Pressure values for different operating points

values into the transfer function Eq. 3.63 and computing the poles yields the pole configuration depicted in **Fig. 3.10**. The complex conjugate poles of the valve $\nu_{v1} = -439.8\frac{1}{s} + 448.7j$ and $\nu_{v1} = -439.8\frac{1}{s} - 448.7j$ are constant. They are not displayed in the graphs.

Fig. 3.10 consists of four two-dimensional graphs. The left two graphs show the system for negatively opened valve, the right side shows the poles for positively opened valve. The upper graphs show the values for real and imaginary part of the poles, while the lower graphs are a side view on the upper diagrams, showing the varying valve opening y_0 on the y-axis. Different symbols for each pole or pole pair allow the mapping of poles in the upper and lower diagram. The legend explains which symbol belongs to which pole or pole pair. The association has been identified by canceling one subsystem at a time and looking which pole disappears. However, it should be pointed out

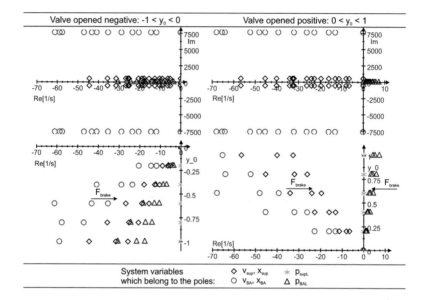

Figure 3.10: Pole configuration for different operating points

that the poles are coupled to each other. This means that changing the stiffness of, for example, the supporting cylinder affects also the poles of the brake actuator as well as the poles associated with the Newton equations of the pistons. Several general observations can be made from Fig. 3.10.

Stability For negative opening of the valve all poles are in the left half plane. The pole associated with the supporting pressure (∗) is, however, very close to zero. This means that the overall behavior is practically integrating. The behavior of the real system is indeed integrating in the sense that the system will not come to rest at a negative valve opening as long as there is brake pressure. As a matter of fact, in the real system the brake force deceases at a negative valve opening until all pressures are zero. But for the linearized system that is a change of the operating point.

The right side shows the characteristics for positive valve opening. The poles associated with the pressure dynamics of the actuator ($\triangle$) are in the right half plane, which means that the system is instable. The pole associated with the pressure in the supporting cylinder ($*$) is also positive, but with a very small value. The integrating behavior which leads to a constant rise in the time domain is superimposed by the instable real pole which results in an exponential rise of the brake force.

Influence of valve opening All poles increase in their absolute values with increased valve opening. The valve is the bottleneck of the system. If it is closed, the flow factors $V_{Q_{PX}}$ become zero. As a result, the change of pressure in one cylinder has no effect on the other anymore.

Influence of brake force The arrows in the graph indicate, that at higher brake forces all poles (except the valve's) move toward zero. This is interesting since one would expect that the poles should move to higher absolute values, since the brake dynamics are observed to be higher for higher brake forces. As we will see in the next section, while the pole wanders to regions of higher time constants, the dominance of that pole is increasing resulting in a steeper slope of the step response.

3.3.1 Comparing dominance of poles

The Litz measure of dominance is a measure to compare the influence of the eigenvalues of a linear system [Lit79], [Föl94]. [Pie94] has applied this technique intensively on hydraulic systems. The basic idea is to determine the factor for each eigenvalue ν_k on the output variable that we are interested in. In our case this is the supporting cylinder pressure, which is proportional to the brake force. The linear system in state space representation is

$$\begin{aligned} \underline{\dot{x}} &= \underline{A}\,\underline{x} + \underline{b}\,u \\ p_{\text{sup}} &= \underline{c}^T \underline{x} \end{aligned} \qquad (3.73)$$

with parameters given in Eq. 3.66. It is possible to find a transformation Matrix $\underline{V}$ with which the system representation is expressed in the Jordan normal form, where all eigenvalues appear on the diagonal of the system matrix. We transform the state vector $\underline{x}$

$$\underline{x} = \underline{V}\,\underline{z} \qquad (3.74)$$

and yield

$$\begin{aligned} \underline{\dot{z}} &= \underline{V}^{-1}\underline{A}\,\underline{V}\,\underline{z} + \underline{V}^{-1}\underline{b}\,u \\ p_{\text{sup}} &= \underline{c}^T \underline{V}\,\underline{z} \end{aligned} \qquad (3.75)$$

for the transformed but equivalent system representation. It follows from Eq. 3.67, that the step response for the system is

$$h(t) = p_0 + \sum_{k=1}^{6} p_k e^{\nu_k t} \qquad (3.76)$$

Since the overall step response to an excitation by the input signal u is the sum of all subsystem's step responses, the factor p_k is a measure for the dominance of each pole ν_k. It can be shown that p_k can be expressed as:

$$p_k = \frac{[\underline{c}^T \underline{V}]_k [\underline{V}^{-1} \underline{b}]_k}{\nu_k} \qquad (3.77)$$

Therefore the Litz measure of pole dominance D_k is defined as

$$D_k = \left| \frac{[\underline{c}^T \underline{V}]_k [\underline{V}^{-1} \underline{B}]_k}{\nu_k} \right| \qquad (3.78)$$

The SEHB system is a single input – single output system. Therefore the input and output variables do not necessarily have to be normalized to appropriate values. However, for the following results, all variables were normalized. The values for normalization were read from nonlinear simulation results for a braking with constant positive valve opening.

$\Delta p_{\text{sup,L}} = 2.97 \; bar \qquad \Delta p_{\text{BA,L}} = 1.47 \; bar \qquad x_{\text{BA}} = 0.0488 \; mm$
$\Delta x_{\text{sup}} = 0.620 \; mm \qquad \Delta y = 100 \; \% \qquad \Delta u = 10 \; V$

The graph in **Fig. 3.11** shows the dominance measure for the six eigenvalues for negative and positive valve opening at three different supporting pressure levels. The complex conjugate eigenvalues are each represented by just one value. One can see that the poles associated with the pressure equations are significantly more dominant than the poles of the Newton equations for both, negative and positive valve opening. It should be pointed out, that the Litz measure of dominance mathematically does not make a difference between instable and stable poles. It is clear, that instable poles are always dominant.

Comparing the poles of the Newton equations, the supporting cylinder is more dominant. The reason is that the capacity associated with the supporting cylinder C_{sup} is parameterized higher due to the influence of the hoses between supporting cylinder and valve block.

Summarizing the results, it seems to be sensible to simplify the dynamic system by crossing out the Newton equations. The simplified linearized model will be derived in the next section.

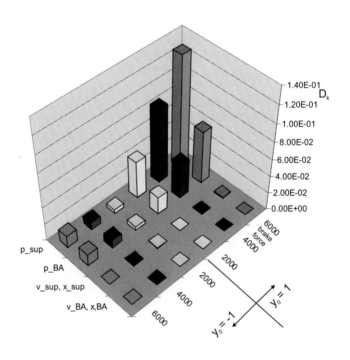

Figure 3.11: Pole dominance measure for different operating points

3.4 Simplified linear model

The reason for neglecting the Newton equations of brake actuator and supporting cylinder pistons has been derived analytically in the previous section. One can also argue descriptively, that acceleration forces are not influential compared to the active forces because the pistons of actuator and supporting cylinder are light-weight.

Deriving the system equation for the simplified linearized model is analogous to the proceeding in section 3.2. After setting up the dynamic equations of all subsystems, they are combined with the help of the Laplace transformation. The only equations to be changed are the dynamic equations of the pistons. The differential equation of the mechanical compression of the brake actuator Eq. 3.32 is simplified by setting all derivatives of the actuator position to zero:

$$0 = A_{BA}\Delta p_{BA_L} - c_{cal}\Delta x_{BA} \tag{3.79}$$

The supporting cylinder differential equation of movement Eq. 3.60 is simplified in the same way:

$$0 = -A_{sup}\Delta p_{sup_L} - c_{s_{sup}}\Delta x_{sup} + \frac{2\mu}{i_L}F_N \tag{3.80}$$

The linearization of the valve flow is given by Eq. 3.7. The valve spool dynamics is given by Eq. 3.10. The pressure build-up equation of the brake actuator is Eq. 3.23 for increasing brake force (Case A) and Eq. 3.28 for decreasing brake force (Case B). The pressure build-up of the supporting cylinder is Eq. 3.50 for increasing brake force (Case A) and Eq. 3.55 for decreasing brake force (Case B).

The transfer function of the simplified linearized system without valve is obtained by applying the Laplace transformation on the differential equations Eqs. 3.23, 3.28, 3.79, 3.50, 3.55 and 3.80. Then the equations can be combined to eliminate all variables except for input $\mathcal{L}(\Delta y)$ and output $\mathcal{L}(\Delta p_{sup_l})$. The initial values of the Laplace transformed variables can be set to zero. The open loop system has four poles: Two for the valve and two for the hydraulic-mechanic system. The transfer function of the hydraulic mechanic system yields:

$$\frac{\mathcal{L}(\Delta p_{\text{sup}_l})}{\mathcal{L}(\Delta y)} = \frac{a_1 s + a_0 s}{b_2 s^2 + b_1 s + b_0} \qquad (3.81)$$

with a_1, a_0, b_2, b_1, b_0 for positive valve opening:

$$\begin{aligned}
a_1 &= V_{\text{sup}_H} C_{\text{BA}_A} C_{\text{sup}}(1+\alpha^3)\big[-K_{\text{sup}_y} i_L c_{\text{s}_{\text{sup}}}(d_1+d_2) + \\
&\quad + 2c_{\text{cal}} K_{\text{sup}_v} \mu A_{\text{BA}} V_{\text{BA}_H} K_y\big] & (3.82) \\
a_0 &= -V_{\text{sup}_H} C_{\text{BA}_A} c_{\text{s}_{\text{sup}}} c_{\text{cal}} i_L \big[C_{\text{sup}} K_{\text{sup}_y}(1+\alpha^3) + \\
&\quad + V_{\text{QPBA}_A} V_{\text{BA}_H} K_y \big] & (3.83) \\
b_2 &= i_L C_{\text{BA}_A} C_{\text{sup}}(1+\alpha^3)(d_1+d_2)(d_3+d_4) & (3.84) \\
b_1 &= i_L C_{\text{BA}_A} C_{\text{sup}}(1+\alpha^3)((d_1+d_2)c_{\text{s}_{\text{sup}}} + (d_3+d_4)c_{\text{cal}}) - \\
&\quad - 2V_{\text{sup}_H} c_{\text{cal}} K_{\text{sup}_v} \mu A_{\text{BA}} V_{\text{BA}_H} V_{\text{QPBA}_A}(1+\alpha^3) & (3.85) \\
b_0 &= i_L c_{\text{s}_{\text{sup}}} c_{\text{cal}}\big[C_{\text{BA}_A} C_{\text{sup}}(1+\alpha^3) - \\
&\quad - V_{\text{sup}_H} V_{\text{BA}_H} V_{\text{QPBA}_A}^2\big] & (3.86)
\end{aligned}$$

and for negative valve opening:

$$\begin{aligned}
a_1 &= V_{\text{sup}_H} C_{\text{BA}_B} C_{\text{sup}}(1+\alpha^3)\big[-K_{\text{sup}_y} i_L c_{\text{s}_{\text{sup}}}(d_1+d_2) + \\
&\quad + 2c_{\text{cal}} K_{\text{sup}_v} \mu A_{\text{BA}} V_{\text{BA}_H} K_y\big] & (3.87) \\
a_0 &= -V_{\text{sup}_H} C_{\text{BA}_B} c_{\text{s}_{\text{sup}}} c_{\text{cal}} i_L \big[C_{\text{sup}} K_{\text{sup}_y}(1+\alpha^3) + \\
&\quad + \alpha^2 V_{\text{QPBA}_B} V_{\text{BA}_H} K_y \big] & (3.88) \\
b_2 &= i_L C_{\text{BA}_B} C_{\text{sup}}(1+\alpha^3)(d_1+d_2)(d_3+d_4) & (3.89) \\
b_1 &= i_L C_{\text{BA}_B} C_{\text{sup}}(1+\alpha^3)\big[(d_1+d_2)c_{\text{s}_{\text{sup}}} + (d_3+d_4)c_{\text{cal}}\big] - \\
&\quad - 2C_{\text{sup}} V_{\text{sup}_H} c_{\text{cal}} K_{\text{sup}_v} \mu A_{\text{BA}} V_{\text{BA}_H} \alpha V_{\text{QPBA}_B}(1+\alpha^3) & (3.90) \\
b_0 &= i_L c_{\text{s}_{\text{sup}}} c_{\text{cal}}\big[C_{\text{BA}_B} C_{\text{sup}}(1+\alpha^3) - \\
&\quad - \alpha^3 V_{\text{sup}_H} V_{\text{BA}_H} V_{\text{QPBA}_B}^2\big] & (3.91)
\end{aligned}$$

with d_1, d_2, d_3, d_4 defined as:

$$d_1 = c_{\text{cal}} T_{\text{BA}_\text{H}} \tag{3.92}$$
$$d_2 = V_{\text{BA}_\text{H}} K_{\text{vBA}} A_{\text{BA}} \tag{3.93}$$
$$d_3 = c_{\text{S}_{\text{sup}}} T_{\text{sup}_\text{H}} \tag{3.94}$$
$$d_4 = V_{\text{sup}_\text{H}} K_{\text{sup}_\text{v}} A_{\text{sup}} \tag{3.95}$$
$$\tag{3.96}$$

The signal flow diagram of the simplified linear SEHB model is shown in **Fig. 3.12**. By combining gain elements in Fig. 3.12, a condensed structure of the simplified system is depicted in **Fig. 3.13**. It shows that the system actually has only one feed-back loop going through a PPT_1 element and a first order lag element PT_1. The PPT_1 element consists of a PT_1 element and a parallel direct feed-through. The DT_1 element is the combination of pressure build-up of the brake actuator (PT_1) and direct feed-through representing the mechanical link between pressure build-up in actuator and supporting cylinder. The pressure build-up of the supporting cylinder is represented by the subsequent PT_1 element. The outer loop from supporting pressure to pressure build-up of brake actuator causes the self-(de)energization process, which was statically described by Fig. 2.2.

Most of the parameters of the open loop system are constant. Only a few are state dependent:

- Supporting pressure p_{sup}
- Friction coefficient μ
- Transmission factor i_L

It is important for the control synthesis to take these factors into account. The supporting pressure range is naturally very wide, since

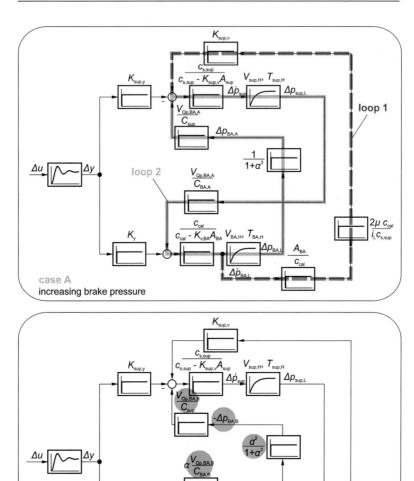

Figure 3.12: Complete signal flow diagram for increasing and decreasing brake force

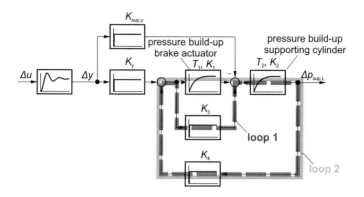

Figure 3.13: Structure of simplified linear system

it is directly connected to the brake force, which is the control variable. The friction coefficient varies typically between 0.3 to 0.4 but can also reach values below 0.15 and above 0.5. It can be calculated using the pressure signals. The mechanical transmission factor varies with the caliper movement, because the angle between the axis of the supporting cylinder and the axis which goes through the center point of caliper rotation and brake force impact point changes. If the caliper movement is measured or estimated, this effect can be taken into account. Using a radial guidance as for the prototypes discussed later in this thesis, the effect of a changing i_L is small compared to the effect of changing p_sup or μ.

3.4.1 Analysis of pole configuration

Since the dynamic elements in Fig. 3.13 are in series, both poles of the hydraulic-mechanic systems must either be positive or negative. It is not possible that only one pole is positive, while the other is negative.

This is interesting, since it was not apparent from the original flow scheme in Fig. 3.7 and Fig. 3.8. The denominator roots of the transfer function Eq. 3.81, the poles, can be calculated analytically:

$$\begin{aligned} 0 &= b_2 s^2 + b_1 s + b_0 \\ \Leftrightarrow s_{1;2} &= -\frac{b_1}{2b_2} \pm \sqrt{\left(\frac{b_1}{2b_2}\right)^2 - \frac{b_0}{b_2}} \end{aligned} \qquad (3.97)$$

Using the pressure relations Eq. 3.69 – 3.72, the poles of this system can be represented analytically dependent on the operating point, defined by brake force and valve opening. **Fig. 3.14** plots the real value of the pole on the z-axis against valve opening and brake force on the basis of the parameter values given in table 3.1.

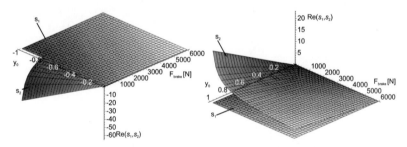

Figure 3.14: Dependency of location of (real) poles (s_1, s_2) on valve opening and brake force

As expected, the poles are similar to the complex linearized model in Fig. 3.10. Both poles are real and positive for positive valve opening and negative for negative valve opening. One of the poles is very close to zero. In section 2.3, Eq. 2.21 describes the condition for self-energization. If the factor $q = \frac{2 A_{\text{BA}} \mu}{i_L A_{\text{sup}}}$ is > 1, self-energization occurs.

Mathematical Model and Control Design   page 61

This criterion must somehow be contained in Eq. 3.97, but it is not possible to simply extract it. The value for q on the basis of table 3.1 is $q = 6.42\mu$. This means that at $\mu > 0.156$ self-energization occurs. **Fig. 3.15** shows the pole locations for $F_{\text{brake}} = 2000\ N$ and $y_0 = 1$ in dependence on the friction coefficient.

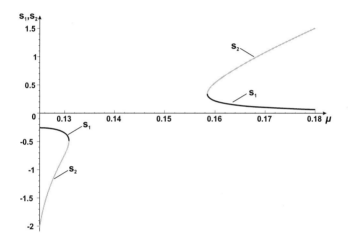

Figure 3.15: Dependence of (real) pole values s_1, s_2 on μ

Eq. 3.97 delivers real values for s_1 and s_2 except for a small region $0.131 < \mu < 0.158$, where they become complex. This means that the dynamics of self-energization is definitely instable and non-oscillating for higher friction coefficients $\mu > 0.158$ because the poles are real positive. The dynamics of self-energization is monotonous rising with rising friction coefficient. The criterion of self-energization ($\mu = 0.156$) lies in the area, where the open loop poles are complex. This means that at $\mu < 0.158$ brake oscillations should be expected. This is an important result to estimate the risk of brake oscillations at low friction coefficients. Lower friction coefficients are technically not relevant, but one can see that the open loop system becomes stable because the poles s_1 and s_2 are real negative.

3.5 Closed loop dynamics

The previous sections mathematically derived and analyzed the open loop system. The results show the open loop instability for positive valve opening. This section analyzes the closed loop dynamics using the simplified model with a second order lag dynamics of the valve and a proportional feedback. The obtained system is of the fourth order.

The focus is on the question, how the proportional gain has to be adapted to the system state to guarantee stability. The friction coefficient μ is uncertain and changes during operation. Also the operating point (brake force and valve opening) varies widely during braking. The goal is to find a functional relation of the optimum proportional gain dependent on the actual brake force and friction coefficient. Unfortunately, the system poles cannot be calculated analytically. However, the pole diagram gives some insight about the influence of controller gain and changes of operating point on the system's stability. **Fig. 3.16** shows the pole configuration for positive valve opening (increasing brake force) in dependence on the controller gain. The diagram is drawn for a brake force of 2000 N and full opening of the valve.

The upper diagram (top view) shows real and imaginary part of the poles. The arrows indicate the movement of poles due to a growing controller gain. The lower diagram (side view) shows the movement of the real part dependent on controller gain. The two poles of the valve are easily identified. They are conjugate complex and move

Mathematical Model and Control Design

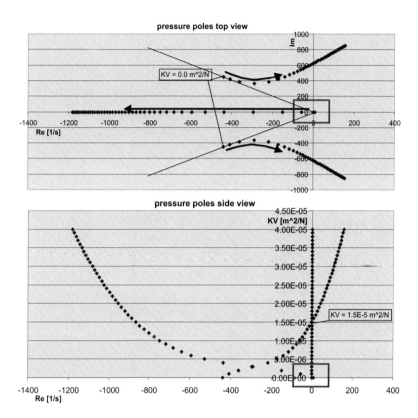

Figure 3.16: Pole movement due to change of controller gain
(positive valve opening)

toward the right half plane for increasing controller gain, which they reach for a controller gain of $K_V = 1.5 \cdot 10^{-5}\ \frac{m^2}{N}$. The controller gain transforms the pressure signal into a control output, the valve input signal u. The valve is fully opened for a control output $u = 1$.

The angle β between pole vector and real axis is a measure for the pole damping $D = \cos\beta$, **Fig. 3.17**. For a damping of $D < \frac{1}{\sqrt{2}} \approx 0.71$ the step response is aperiodic, [Abe06]. Smaller damping in-

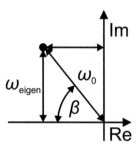

Figure 3.17: Characteristic values of complex conjugate poles

evitably leads to oscillation. The valve has been parameterized with a characteristic frequency of $f_{\text{eigen}} = \frac{\omega_0}{2\pi}\sqrt{1-D^2} = 55\ Hz$ and a damping of 0.7. **Fig. 3.16** delivers the important result, that oscillation is to be expected for this system for any gain. None of the valve poles is within the 45°-line which marks the aperiodic case. This leads to the conclusion that for SEHB, the valve damping is more crucial than its characteristic frequency.

The two pressure poles have no imaginary part for most controller gains. One of them tends hard to the left, the other seems to remain around zero for increasing gain. **Fig. 3.18** is a more detailed view of the boxed area in Fig. 3.16. For very low controller gain, some poles are complex conjugate. One can also see that there is a lower limit for the controller gain at $K_V = 1.28 \cdot 10^{-7}\ \frac{m^2}{N}$. Below $K_V = 1.28 \cdot 10^{-7}\ \frac{m^2}{N}$ the system has conjugate complex instable poles. For values $1.28 \cdot 10^{-7}\ \frac{m^2}{N} < K_V < 1.41 \cdot 10^{-7}\ \frac{m^2}{N}$ the system is theoretically oscillatory stable but also very slow. After both poles meet, they

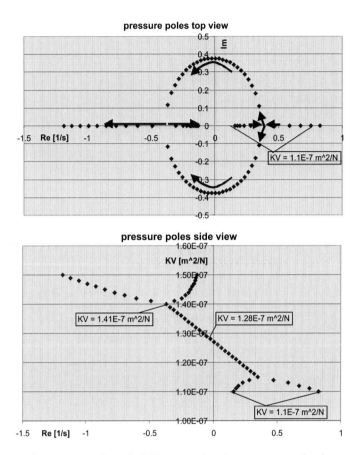

Figure 3.18: Detail of Fig. 3.16 showing pressure poles for small controller gains

leave in opposite directions. The pole leaving to the right potentially destabilizes the system but it does not cause oscillations like the valve poles.

One might expect that the controller gain for positive valve opening is also satisfactory for negative valve opening. But actually the system dynamics is faster for decreasing brake force. The pole wandering is plotted in **Fig. 3.19.** for the same operating point and range of control gain as in Fig. 3.16 but negative valve opening.

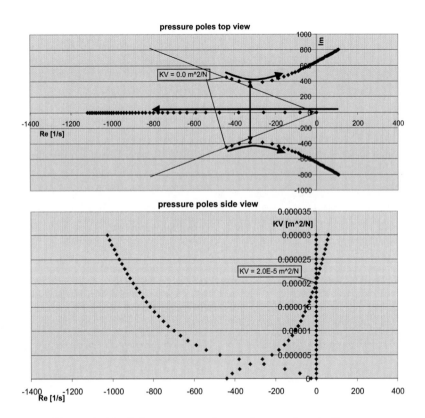

Figure 3.19: Pole movement due to change of controller gain (negative valve opening)

Both cases, positive and negative valve opening, lead to similar pole configurations. An important difference is that the valve poles of Fig. 3.19 are slightly more distant from the 45°-degree line. This means that their damping is lower and oscillation is higher. The conclusion is that the maximum gain is limited by the closed loop dynamics for negative valve opening.

3.6 Adaptive proportional control

According to the results of the previous sections, the controller gain must be adapted to system state parameters. The main factors are those which directly influence the dynamics of self-energization. The conclusion of section 3.4 is that of the state dependent parameters, the actual brake force and friction coefficient should be taken into account. If the mechanical transmission factor i_L is state dependent it should be taken into account as well.

To demonstrate the effectiveness of the adaptive control and for verification of the results of this chapter, this section presents simulation results of a simplified nonlinear hydraulic system simulation using adaptive proportional gain.

The scheme in **Fig. 3.20** represents the nonlinear model for system simulation in DSHplus. The model is parameterized with the values of table 3.1 except for the valve damping, which is set to $D_V = 0.8$. The springs are included in the cylinder models. Internal friction in the cylinders is set to zero for the purpose of demonstration. With friction in the cylinders, the system has less tendency to oscillate. The pressure lines model the capacity according to their volume combined with a state dependent correction of the bulk modulus.

The linearized model can be utilized to find parameters to adapt to the changing system dynamics. The brake force should not overshoot

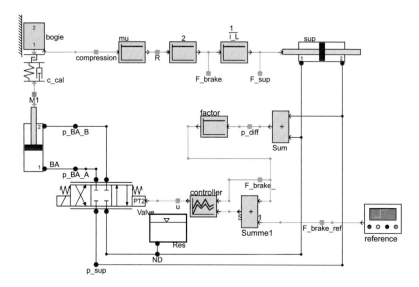

Figure 3.20: Simplified simulation scheme for nonlinear system simulation in DSHplus

or vibrate in closed loop control. Therefore a criterion for calculating the controller gain could be to set the damping of the valve poles to 0.7. As explained in the previous section, to achieve this the open loop valve poles have to have a damping greater than 0.7. The open loop valve damping coefficient therefore is defined as $D_V = 0.8$ for calculating the controller map. The controller map is parameterized according to **Fig. 3.21**. Two curves have been identified for different control design criteria $D_1 = 0.7$ and $D_2 = 0.55$.

Fig. 3.22 shows the simulation results for step responses from 1 to 12 kN using the controller maps of Fig. 3.21. As expected, the simulation using a controller map for $D_1 = 0.7$ shows no overshoot

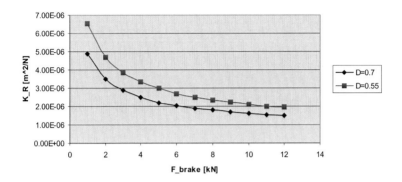

Figure 3.21: Controler map in dependency of actual brake force

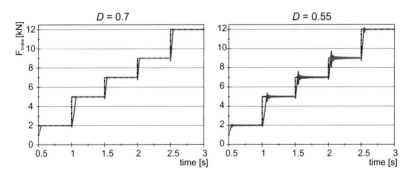

Figure 3.22: Simulation results using adaptive proportional control

or oscillation. The control performance using the controller map for $D_2 = 0.55$ is, however, not acceptable. This example illustrates the importance of adapting the proportional gain. The compromise in control performance is huge if only a constant feedback gain is parameterized for the whole operating range. The linearized model can be used to calculate the controller map. Because of cylinder friction which has a damping effect in the real process, the obtained map represents a lower boundary for the control gain of the real system. The real control gain can be greater by a factor which has to be proved in experiments.

Chapter 4

Hydraulic-mechanical design of SEHB

All design requirements of the self-energizing brake have to be fulfilled by the hydraulic-mechanical design. This encompasses operational safety, dynamics and control performance and comfort as well as energy efficiency. This chapter discusses the main design requirements and basic design rules.

The following section 4.1 discusses design requirements for SEHB which are partly in conflict.

It is often difficult to distinguish between mechanical and hydraulic design because it is clear that every hydraulic component is also designed mechanically. Since the force transmission and control is done by hydraulic elements, the hydraulic design is derived more or less directly from functional requirements. The aim of section 4.2 is to give simple calculation formulas which are related to the brake functionality. In contrast, most of the mechanical design depends on operating conditions, installation space and on experience with certain mechanical elements like bearings, slider-mechanism etc. in the specific field of application. Section 4.3 systematically discusses different solutions for major aspects of the mechanical design. **Fig. 4.1** gives an overview of the major hydraulic and mechanical design elements which are explained in the following pages.

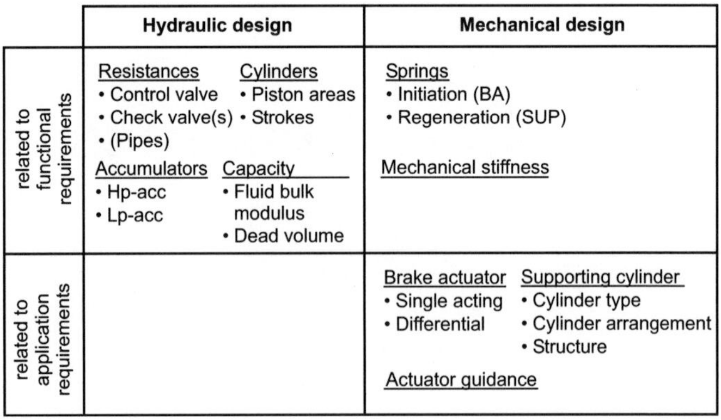

	Hydraulic design	Mechanical design
related to functional requirements	Resistances Cylinders • Control valve • Piston areas • Check valve(s) • Strokes • (Pipes) Accumulators Capacity • Hp-acc • Fluid bulk • Lp-acc modulus • Dead volume	Springs • Initiation (BA) • Regeneration (SUP) Mechanical stiffness
related to application requirements		Brake actuator Supporting cylinder • Single acting • Cylinder type • Differential • Cylinder arrangement • Structure Actuator guidance

Figure 4.1: Hydraulic and mechanical aspects of SEHB

4.1 Design requirements

Some design goals of SEHB concern basic requirements that every brake should fulfill. Other design goals concern boundary conditions of the target application.

4.1.1 General design requirements

General requirements are:

- Safe operation at a low friction coefficient
- Good controllability
- Energy efficiency
- High dynamics
- Compact design

In most cases a brake is safety relevant. To assure safe operation for all circumstances e.g. icy or wet disks and brake pads, the SEHB should be designed for friction coefficients down to 0.1. Of course, the development of friction materials aims to achieve a constant friction coefficient over wide operating conditions. One reason is because the wheel slide control algorithms depend on a predictable brake behavior. Therefore regulations are quite strong for friction material to maintain a constant average friction coefficient [uic04].

Controllability is guaranteed by a robust input-output behavior of the brake. The brake should deliver smooth and predictable deceleration momentum and it should not lock up unexpectedly. The brake dynamics are analyzed theoretically in chapter 3 and is demonstrated in experiment in chapter 6. To enhance the controllability, special attention has to be given on the valve concept in see chapter 5.

Energy efficiency has been a major motivation for developing the SEHB. Electric energy is only needed for the control unit and to drive the solenoids of the control valve(s). To power the brake, the vehicle's kinetic and potential energy is extracted from the braking process by the supporting cylinder. The stiffness of the hydraulic and mechanical system is responsible for the energy consumption to achieve a certain brake force. The higher the stiffness, the less volume is needed to produce the desired brake force. But this is a design conflict between efficiency and controllability because a higher stiffness also results in faster pressure build-up. The valves must be designed appropriately concerning nominal flow and dynamics to control very small flows. Changes of brake disk thickness and shape due to irregular temperature and wear distribution lead to higher brake force vibration, which reduces the brake comfort. **Fig. 4.2** depicts this design conflict. Although dynamics may be less with increased comfort

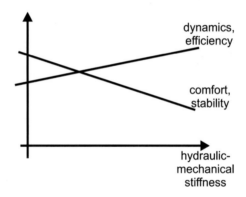

Figure 4.2: Design conflict between pressure efficiency and brake comfort

and stability, the reader should be reminded, that high dynamics is a relative term. Train brakes, which are conventionally air brakes, have comparatively low stiffness and dynamics but yet enough for more or less efficient wheel-slip control. The SEHB has higher dynamics due to the stiffness of the pressure fluid.

Compact design is always relevant, especially for mobile machines, where the space is needed for people or transportation of goods and additional weight leads to less efficient service. The SEHB can be designed to be lightweight compared to conventional train brakes because it does not require a power pack and pressure lines through the whole train. Its reservoirs are small compared to the ones used presently.

4.1.2 Application specific design requirements

The SEHB was first developed for a train application. Besides the general requirements, some special requirements have to be met by a train brake:

- Fail-closed concept
- Inexhaustibility of brake power
- Active setting of clearance
- Wheel slide control
- Parking brake
- Rough environmental conditions / impact-endurance
- Load adaptation of safety brake

The SEHB concept is able to address all of the above requirements. The fail-closed concept, as already mentioned above, is realized by spring centered fail-safe positions of the valves. The inexhaustibility is important for train brakes since the air compressor is mostly centralized in the locomotive. According to the regulations, pneumatic brakes meet the requirement of inexhaustibility, when enough air is stored in decentralized reservoirs to fully compress and release the brake 5 times. Translated directly into the SEHB concept, the supporting cylinder needs to have enough stroke to supply for 5 brakings. This has been the objective of the SEHB design so far, but it disregards the fact that energy supply of the SEHB is not centralized in the first place. Each caliper has its own power supply. The danger of a single failure that affects all brakes' supply like a separation of train cars, which cuts the main pressure line of an air brake system, does not exist. The advantage of SEHB is that it fundamentally is inexhaustible. More about this aspect can be read in [LS07].

The active setting of 1 mm brake clearance on each side prevents contact between brake pads and disk which could lead to heating up and glazing of the brake pads. Conventionally the brake is hinged on levers attached to the bogie which gives additional axial flexibility. This is needed because the brake disk is fixed on the axle and side impacts cause axial displacement. The active clearance also avoids tolerance conflicts.

Wheel slide control systems are important for trains because they guarantee the specified braking distances, which determine the possible capacity utilization of tracks. Conventionally the wheel slide control unit is separate from the brake control unit with a special approved control implementation and separate valves. It is possible to develop analogous concepts for the SEHB, although it would be much more convenient to use an adaptation of the set-value for wheel slip control.

Load adaptation is also needed for the safety brake to guarantee braking distances. A fully loaded train needs more deceleration force than an empty train. An idea for a mechanical acting load adaptation in combination with hydro-mechanical control of the deceleration force instead of electrical control is presented in [LS07, SLS06a].

Parking brakes must safely halt the vehicle for an indefinite time. If the vehicle stands on a slope, the self-energization is able to lock the wheel. The sealings of the supporting cylinder may however be the weak point as well as any internal leakage. If the supporting cylinder reaches its end position, the brake force will cease. One solution for the parking brake is to design the spring in the brake actuator to be strong enough to function as a parking brake. Another solution explained in [LS07] is to use seat type valves to connect a brake pressure accumulator tightly to both sides of the brake actuator, [SLS06b].

The rough environmental conditions require an appropriate mechanical design, which has not yet been considered in this phase of research.

4.2 Hydraulics

The functionality of SEHB is realized by the hydraulic-mechanical scheme depicted in Fig. 2.9. The basic hydraulic components included in this scheme are:

- supporting cylinder
- brake actuator
- control valve (may be composed of several valves)
- two suction check valves
- two high pressure check valves
- high pressure accumulator
- reservoir

The following subsections are a guideline for the design of these components.

4.2.1 Cylinders

The precondition for self-energization is that the supporting pressure generated by a braking force exceeds the pressure which is needed to cause this braking force. Mathematically this precondition is expressed by $q > 1$, with q being the ratio between added pressure force component F_{inc} and brake force F_{brake}, see Eq. 2.3. Factor q has been derived for an ideal SEHB without friction in Eq. 2.24. The precondition then is:

$$\frac{2A_{\text{BA}}\mu_{\min}}{i_{\text{L}}A_{\sup}} > 1 \tag{4.1}$$

Spring forces and piston friction, however, have a significant influence on the precondition of self-energization especially at low brake forces. The spring force of the initialization spring in the actuator $F_{\text{s}_{\text{BA}}}$ initiates and supports the self-energization. The force of the retraction spring $F_{\text{s}_{\sup}}$ and all friction forces $F_{\text{f}_{\text{BA}}}, F_{\text{f}_{\sup}}$ oppose this process. Including these forces into the ratio q between added pressure force component F_{inc} and brake force F_{brake} yields:

$$q = \frac{F_{\text{inc}}}{F_{\text{brake}}} = \frac{2\mu_{\min}(p_{\sup}A_{\text{BA}} + F_{\text{s}_{\text{BA}}} - F_{\text{f}_{\text{BA}}})}{i_{\text{L}}(p_{\sup}A_{\sup} + F_{\text{s}_{\sup}} + F_{\text{f}_{\sup}})} \tag{4.2}$$

This equation shows that at low brake forces the initialization spring has to be strong enough to overcome the friction forces. To be able to start the process of self-energization, the initialization spring has to apply a force of

$$F_{\text{s}_{\text{BA}}} = \frac{i_{\text{L}}(F_{\text{s}_{\sup}} + F_{\text{f}_{\sup}})}{2\mu_{\min}} + F_{\text{f}_{\text{BA}}} \tag{4.3}$$

4.2.2 Valves

The central valve of SEHB is the control valve. The control valve must be very tight in the closed position. In case of lifted brake pads, the piston ring side is pressurized by the initialization spring. Leakage leads to a movement of the pads onto the brake disk. In case of braking, leakage from the high pressure side to the reservoir empties the supporting cylinder faster, which must also be avoided. As has been shown in chapter 3, the closed loop gain is brake force dependent. To achieve high dynamics for low brake forces, a high flow gain is desired. As the brake dynamics exponentially increase for higher brake forces, a low flow gain is needed. Since no characteristic dynamic value for the brake has already been defined, it is not yet possible to quantify the needed flow gain dependent of the actual brake force. The valve should ideally have a high proportional resolution to avoid pressure steps. A relation between a minimum valve opening and the resulting minimum pressure step has been derived in [LSM08].

The suction and the high pressure check valve are combined to realize the function of a hydraulic rectifier. The suction valves should be chosen to be large enough to avoid cavitation. During braking, especially in the beginning when the high pressure accumulator is charged, the supporting cylinder moves a short distance but quickly. The pressure drop across the suction check valves can cause cavitation which may damage the valve's seat. Also, pressures below atmospheric pressure in the supporting cylinder may lead to suction of air through the piston

4.2.3 Accumulators

During the beginning of a braking the high pressure accumulator stores the energy needed for retraction of the brake actuator. The clearance is 1 mm on each side, so the volume needed for retraction V_{acc} is:

$$V_{\text{acc}} = \alpha A_{\text{BA}} \cdot 2\ mm \tag{4.4}$$

The pressure level at which this volume must be provided, depends on the initialization spring force F_{sBA}, the friction force F_{fBA} and the pressure drop over the valve, which depends on the desired retraction velocity.

The low pressure reservoir provides the differential volume of the brake actuator. It can be pressurized with a small pressure to avoid too low values of the oil bulk modulus especially at low brake forces. A pressurized reservoir also reduces the risk of sucking air through the piston rod sealing of the low pressure side of the supporting cylinder.

4.3 Mechanics

This section focuses on aspects of the mechanical implementation of the SEHB for a specific application. In the following sections, structural alternatives of different aspects of the mechanical design of SEHB will be discussed for the following parameters:

1. Brake actuator type
   - Fixed caliper or pin slide caliper
   - Retraction of brake pad passive or active
2. Actuator guidance

- Radial support: exact circular guidance
- Linear support: shifting friction radius
- Curve approximation / guidance mechanism

3. Supporting cylinder
    - Double rod cylinder
    - Differential cylinder
    - Arrangements of 2 independent plungers
    - Integrated design of 2 plungers
    - Rotary actuator

4. Supporting cylinder arrangement
    - Alignment of supporting cylinder on vector of brake force
    - Mounting orientation of supporting cylinder

4.3.1 Brake actuator

Two major types of brake actuators can be distinguished: the fixed and the pin-slide caliper, **Fig. 4.3**. The brake disk (3) is axially fixed. The caliper (4) encompasses the bake actuator(s) (2) and applies the compressive forces. While the brake pads (1) are thrusted forward by the brake actuator(s), they are laterally supported in the bracket (5). In the case of the fixed caliper (left in Fig. 4.3), two separate hydraulic plungers with equal piston areas thrust the brake pads from both sides. In this case the bracket guiding the brake pads and the plungers are integrated into one part. Another solution is the pin-slide caliper(right in Fig. 4.3). In this case bracket and caliper are separated. The bracket is connected to the bogie structure. It is responsible for laterally guiding the pads and conducting the brake

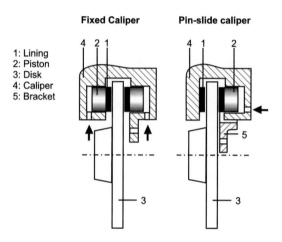

Figure 4.3: Fixed and pin-slide caliper brake

force into the fixed structure. The caliper is guided on pin-slides which are connected to the bracket. Only on one side it features an actuator. When the actuator is pressurized, it pushes both pads symmetrically onto the brake disk, while it centers itself on the brake disk sliding on the pins.

One advantage of a fixed caliper is that it has no moving parts except the pistons. The hydraulic connection can be stiff piping instead of flexible hose. With the concept of SEHB, however, during braking there is always movement between brake actuator and pressure source. Advantages of a pin-slide caliper over a fixed caliper are:

- Reduced installation space
- Reduced weight
- Less external sealings

- Only one contact surface between piston and brake pad reduces heat transfer into fluid

The retraction of the brake piston is necessary to lift the brake pads off the disk. In many brake designs the retraction is realized passively. The slight unevenness of the disk pushes back the brake piston while at the same time it is pulled by the elasticity of the sealing or an extra spring. The gap between the pads and disk of conventional automobile brakes is typically $40 - 60$ μm. For trains it is common practice to have a gap of $2-3$ mm, which is guaranteed by prestressed springs. The reason for the large gap of train brakes is that there is no stiff connection between caliper and wheelset. The brake pads are hanging on links mounted to the bogie which can move vertically or tilt in relation to the wheelset. To prevent frequent or permanent contact between brake pads and disk, which can lead to glazing of the brake lining, the gap is set to a higher value. Therefore active retraction is needed. For the SEHB this can be done by using the pressurized fluid in the high pressure accumulator from the previous braking. **Fig. 4.4** shows the difference between passive and active retraction of the brake piston.

4.3.2 Actuator guidance

A unique characteristic of SEHB is the brake pad movement in direction of the brake force, producing the necessary hydraulic power in the supporting cylinder, which in turn is used for brake actuation. This movement must be guided in some way to avoid an overlap of the brake pads beyond the disk. While it may be appropriate in some applications to allow small overlap, an exact or approximated circular guidance is certainly required in most cases. Two options to realize this guidance are depicted in **Fig. 4.5**

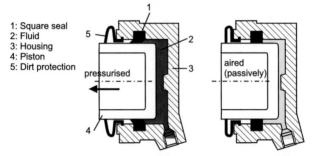

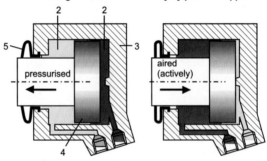

Figure 4.4: Passive and active retraction of single and double acting cylinder

The most obvious way to realize circular guidance is to connect the caliper with a radial bearing to the brake shaft, as shown in the Fig. 4.5 a). The radial bearing appears to be simple at first, but it has some significant disadvantages. First, it produces loss in the drive train. Moreover, it must be designed to be replaceable. In applications where it cannot be mounted on one of the axletree's ends, it needs to be made of two separable parts. The shaft radius of a

Hydraulic-mechanical design of SEHB

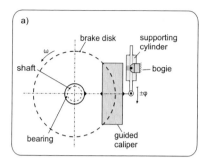 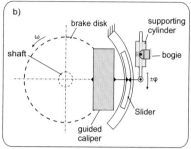

Figure 4.5: Circular guidance through radial bearing (a) or slider (b)

train's wheelset is typically around 180 mm, while for the purpose of the bearing 40 mm would be sufficient. The attachment of the caliper to the wheelset is contrary to today's service procedures. Before changing the wheelset, the brake would have to be disconnected from the shaft. Therefore alternative solutions for circular guidance are necessary. The Fig. 4.5 b) depicts a solution using a slider guided in a circular groove. The suitability of this solution in the rough environmental conditions that a bogie is exposed to is questionable. A third solution using only simple joints is the application of a Watt-I 6 link mechanism, [KPC07]. The basic scheme of a Watt-I linkage, where the first beam (0;6) is fixed, is shown in **Fig. 4.6**.

The solution for radial guidance of lever (4) around a point (P_0) on the line through part (0;6) develops as explained in the following: With part (1) and (3) being straight beams crossing each other in the middle and levers (2), (4) and (5) each half as long as the beams (1) and (3), lever (4) is guided as if rotating around point (P_0), which is the intersection of (0) and (5). This mechanism is applied on SEHB in **Fig. 4.7**. The advantage of this 6 link mechanism is, that no

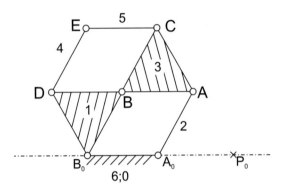

Figure 4.6: Watt-I linkage

machine parts have to be near (P_0). The brake can be completely mounted to the bogie.

4.3.3 Supporting cylinder type

The supporting cylinder design has significant influence on the installation space. **Fig. 4.8** provides an overview of possible configurations for different supporting cylinder types.

To facilitate bi-directional braking, both a double rod or a differential cylinder come into question, see solution (1) to (4) in Fig. 4.8. However, one requirement of brakes for trains is that the braking effect is independent of the vehicle's direction of motion. Therefore, the supporting cylinder must have equal piston areas.

The double acting cylinder can be divided into two plungers as shown in solution (5) – (13). Using plungers offers greater design flexibility, which can be useful when aiming on large scale integration. Plungers are also cost-efficient for mass production. Another advantage is that

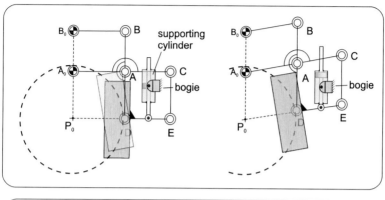
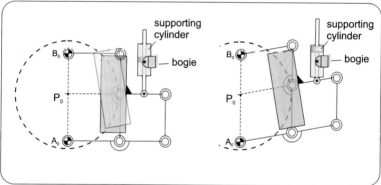

Figure 4.7: Circular guidance of the caliper by a 6 link mechanism

the diameter of a plunger compared to a double rod cylinder with the same piston area is much smaller. Arrangements (11) – (13) should be especially highlighted. Here, the plungers are loosely connected to the caliper. They can be pushed only in one direction of actuation. This means that during braking only one plunger is pressurized while

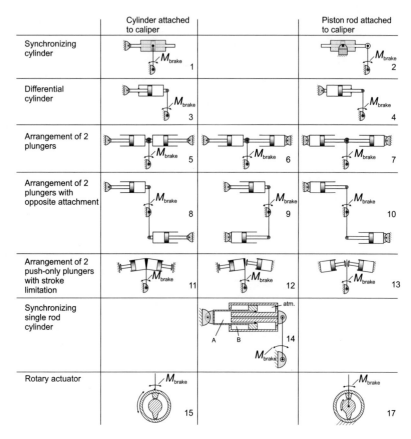

Figure 4.8: Configurations using single or double rod cylinders

the other is limited in its end position. This allows to connect both plungers directly to each other, [SL08]. In all other solutions, where the supporting chambers are kinematically bound, the pressurized chamber has to be separated by check valves from the low pressure chamber, Fig. 2.9. These check valves can be avoided with solutions (11) – (13). Also to initialize the supporting cylinder, in all other solutions a switching valve is needed to bridge the check valves. The supporting cylinder can only move into its center position when both chambers are connected. This switching valve is no longer needed when both plungers are directly connected.

A very interesting solution is the synchronizing single rod cylinder as shown in (14). It looks like a differential cylinder but features two equal piston areas. Pressurizing chamber A leads to a pushing force. Pressurizing the ring area in chamber B leads to a pulling force. This design has significant advantages. It is very compact in length and not much wider than the double rod cylinder. Spherical joints can be mounted on both ends. A drawback from a production viewpoint is that it requires small tolerances. The cylinders have to fit very well into each other.

To complete the systematics of supporting cylinder arrangements, rotary actuators (15) and (16) can be used to support the brake force. At first glance they may facilitate very compact solutions, allowing large scale integration. However, one difficulty is the increased friction compared to a conventional cylinder.

4.3.4 Supporting cylinder arrangement

The arrangement of the supporting cylinder is a central aspect of the design of SEHB. This section discusses the geometric orientation and the cylinder types that can be used.

- Alignment of supporting cylinder on the vector of the brake force

- Pivot point in or adjacent to the brake surface
- Axis of supporting cylinder in line with or placed in parallel to the vector of brake force

• Mounting orientation of supporting cylinder

- Piston rod attached to brake actuator (cylinder fixed)
- Cylinder attached to brake actuator (piston rod fixed)

Alignment

A support of the brake force by the supporting cylinder with a vertical or horizontal displacement leads to stress in the parts between the brake bracket, which holds the brake pads and its links to the wheelset and bogie. This fact is illustrated in **Fig. 4.9**. The brake

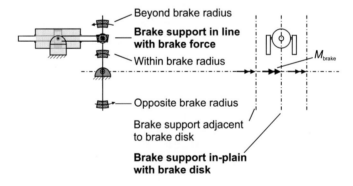

Figure 4.9: Point of application of supporting and brake force

force vector lies in the center of the brake disk tangential to the friction radius. By principle, the supporting cylinder carries only an axial load. Therefore, a vertical displacement between the supporting cylinder and the brake force (see left side in Fig. 4.9) leads to a bending moment in the bracket, which is compensated by additional bearing forces in the joints connecting the bracket to the wheelset and the supporting cylinder. Also flexing stress is caused in the bracket. Both necessitates stronger links and components. A horizontal displacement (right side in Fig. 4.9) adds another bending moment in vertical direction.

The attachment of the supporting cylinder(s) directly onto the guiding bracket of the brake pads would be ideal to minimize flexing stress in the brake.

Mounting orientation

The supporting cylinder should carry only axial forces since lateral forces increase the friction force of the cylinder and cause wear of the guide sleeves. The easiest way to eliminate transverse forces is the use of spherical joints, which can be mounted on the piston rod ends but not on the cylinder. The differential cylinder offers the possibility to be fixed by two spherical joints, one mounted to the piston rod and the other to the bottom end of the cylinder. The double rod cylinder is normally mounted over two linkage stubs attached to both sides of the cylinder. They allow a one-dimensional tilting only and have to be aligned precisely to prevent transverse forces. For the generation of the supporting pressure, it does not make a difference whether the piston rod is attached to the caliper and the cylinder to the bogie or the opposite way. For the mechanical design it makes a difference because the piston rod end and the spherical joint are much smaller than the cylinder. Advantages resulting from the attachment of the cylinder to the caliper are:

- Integrated design possible of supporting cylinder and caliper

- Single serviceable unit
- Simple mechanical connection to bogie
- No hydraulic lines between bogie and caliper

Advantages resulting from the attachment of the piston rod to the caliper are (see upper right of Fig. 4.8):

- Compact caliper
- Hydraulic valves mounted on cylinder are shock protected
- Reduced unsprung mass
- Connection of supporting cylinder to bracket within brake radius less complex

Chapter 5

Valve configurations for pressure control

5.1 Valve arrangements

The basic task of the valve control of SEHB is the control of the pressure difference at the supporting cylinder. It is often stated that spool type valves with some negative overlap of control edges are best suited to achieve fast, smooth and precise control performance [Mur08]. Negative overlap of control edges leads to a smooth transition from the connection of the control port to the pressure source over to the tank. But it also leads to leakage between the pressure source and the tank. Even precisely lapped spool type valves always exhibit leakage in the closed center position due to the radial clearance between spool and sleeve. With a maximum volume of 38 ml in each chamber of the supporting cylinder used as the pressure source of the SEHB prototypes, valves which allow leakage cannot be used. Therefore seat-type valves are preferred for use in SEHB.

Most commercially available seat-type valves are switching valves. Switching valves have no point of equilibrium for the armature between the open and closed position. The magnetic force on the armature rises exponentially with the closing gap. This leads to an optimized switching time and low energy consumption for holding the armature after switching. For some applications, solenoids have been developed which have a continuous current-opening characteristic. This is achieved by shaping the armature in such a way that the gradient of the magnetic force due to change of the armature position is low in the operating point. Seat-type valves, however, have the disadvantage that in closed position a pressure difference between its ports result in a force on the actuation element if the design is not pressure compensated. The pressure force on the actuation element is a state dependent disturbance on the valve opening control because the solenoid current - valve opening characteristic is shifted. To be able to use these valves in closed loop control, pressure and/or position sensors in combination with a strong actuation to compensate pressure forces are needed.

The basic functional scheme of SEHB with a differential cylinder as brake actuator introduced in Chapter 2.4 uses a 4/3-way valve as control element between high and low pressure lines and both sides of the brake actuator. The zero overlapped 4/3 way valve couples four hydraulic resistances in a fixed way. As shown in **Fig. 5.1**, different alternative arrangements of multiple valves can replace that 4/3-way control valve. The four coupled control edges in the 4/3 way valve can be distributed on two 3/2 way valves or four 2/2 way valves. Another option is to connect the ring side of the actuator with high pressure and to control only the pressure on the piston face side, also known as A+E-control [Mur08]. For full versatility and maximum

Valve configurations for pressure control

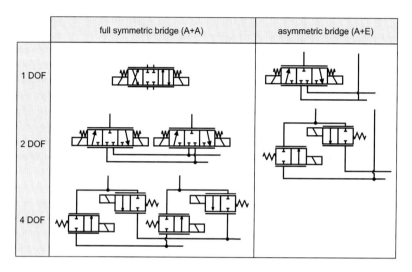

Figure 5.1: Degrees of Freedom (DOF) of different valve arrangements

dynamics the combination of four 2/2 way valves is chosen for the SEHB prototype.

5.2 Fast switching 2/2-valves

Fast switching seat-type valves have some advantages and some disadvantages compared to proportional spool type valves. Their tightness in the closed position and comparatively low costs make them ideal for use in SEHB. A drawback is that the abrupt switching causes pressure steps and oscillation. Every opening of the valve leads to a definite amount of volume transport which leads to a definite pressure rise in the controlled volume. The valve flow as a response to a reference pulse can be characterized by 6 phases ($V_1 - V_6$), as shown in **Fig. 5.2**, [Wen96].

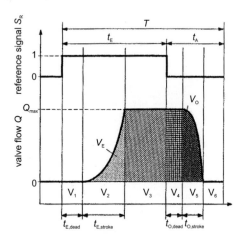

Figure 5.2: Characteristic flow response to a pulse input, [Wen96]

1. The solenoid current builds up during the dead time of valve engagement but is too low to pull the armature against the spring force.

2. The armature is pulled up opening the valve until it hits the upper end.

3. The valve is fully opened. The magnetic flux further builds up until saturation is reached.

4. The magnetic flux dissipates when the supply voltage is cut off and the solenoid is hot-wired over a flyback diode and a

shunt. This continues until magnetic force and spring force are in balance.

5. Magnetic force drops below the spring force. The valve closes until the valve seat is tightly closed.

6. The solenoid current further drops, while the armature is pressed into its seat until the spring force is fully effective.

The following section 5.2.1 provides an overview about literature on hydraulic (pressure) control using switching valves.

5.2.1 Pressure control using fast switching valves

Pressure control using (fast) switching valves is not a common practice for stiff systems. It is widely used in pneumatics though, where compressibility is higher. In hydraulic systems, the discontinuous flow results in a sudden pressure step, which is able to excite higher order dynamics in the structure. Since the 1970's several studies have been made for the use of switching valves to substitute proportional or servo-valves. There have been three main motives for this:

- Cost reduction due to mass production and less demanding production accuracy
- Low sensitivity to contamination compared to servo valve
- Tightness achieved by seat-type design.

Most studies found on closed loop control of hydraulic systems with switching valves focus on position control [HM72], [Ter92], [Kür93], [Pro86], [Bec04], [Fis99], [Wen96]. Some of these, [Kür93], [Bec04], [Fis99] and [Wen96], also cover some aspects of pressure control. As a pressure source they assume a constant pressure system. They all use pulse width modulation (PWM) to produce quasi-continuous flow with the switching element. [Wen96] focuses especially on pressure control with switching valves in automotive applications and analyzes the behavior of different 2/2 and 3/2-way switching valves at different pressure levels and pulse lengths. The automotive anti-lock brake system (ABS) is a famous application for switching valves in hydraulic pressure control. Although, in most anti-lock brake systems, it is not the pressure that is the control variable but the slip. Since the nominal flow of the anti-lock system's valves is similar to that required for SEHB, such valves are used for the first prototype

Proportionally acting control is the easiest and best strategy for pressure control of closed cavities [Wen96]. No relevant work in literature was found on nonlinear control methods like dead-beat control which could specifically make use of the nonlinear behavior of the switching valve. Instead PWM was used in the literature found to emulate continuous behavior. However, a certain control deviation cannot be eliminated even with PWM because of the system immanent pressure step size, which is discussed later. A deadband in the control loop prevents oscillation around the set value, **Fig. 5.3**

By a specific short pulsing of the switching valve, achieving a partial opening, it is possible to increase maximum bandwidth and control resolution. This presumes a very accurate model of the switching valve. [Kür93] shows how simplifications or uncertainties in modeling influences the control result.

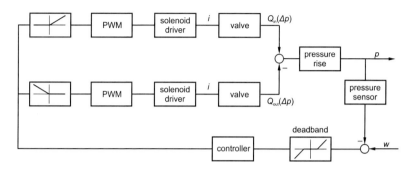

Figure 5.3: Control strategy for pressure control with switching valves

5.2.2 Anti-lock brake system switching valves – characteristics and driver electronics

This section presents the switching valves which were used for the first prototype tests and explains the simple electronic switch used for driving the solenoid.

The valves used for the first prototype on the basis of an automotive brake caliper are switching seat valves with solenoid actuation, see **Fig. 5.4**. They are designed as cartridges which are pressed into a special boring in a block of aluminum forging alloy. The cutting edges cut into the material and create tight connections between cartridge and boring separating inlet and outlet of the valve as shown on the left side of Fig. 5.4. The armature with the seat element connected to its lower end is encapsulated in a cylindrical housing. The solenoid is mounted over it and acts through the housing.

The valves are available in fail-open and fail-closed design. The inlet and outlet of the valves are such that the flow supports the fail-safe

Figure 5.4: Anti-lock brake system switching valves from Continental Teves

position of the valve. For a fail-closed valve the inlet is the horizontal upper port, for a fail-open valve the inlet is the vertical lower port.

The switching times of fail-open and fail-closed valves can be measured indirectly by measuring the spool current. **Fig. 5.5** shows the electronic switch used for actuating the valve. If the base of the NPN-dotted bipolar junction transistor of type TIP 142 is switched to positive by the real-time control system, the connection between collector and emitter becomes conductive. Current builds up through the coil, which is in series with a measuring shunt. A differential amplifier circuit delivers the difference of voltage U_s before and after the shunt R_s as a measure of the coil current according to Ohm's Law. The values for resistances and supply voltage are:

$R_{\text{fail-openvalve}} = 6.6 \ \Omega \quad R_{\text{fail-closedvalve}} = 6.6 \ \Omega \quad R_1 = R_3 = 10 \ k\Omega$
$R_2 = R_4 = 20 \ k\Omega \quad R_s = 0.9 \ \Omega \quad u_0 = 9 \ V$

Valve configurations for pressure control

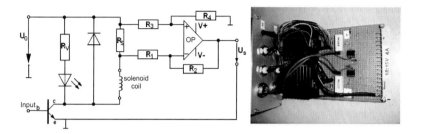

Figure 5.5: Spool driver electronics for anti-lock system valves of first Prototype

The differential amplifier amplifies the voltage difference by factor $\frac{R_2}{R_1} = \frac{R_4}{R_3} = 2$. Therefore the coil current is calculated by

$$i_{\text{coil}} = \frac{u_s}{2R_s} \qquad (5.1)$$

A flyback diode is placed parallel to the solenoid to eliminate the flyback when the supply voltage is suddenly switched off. It provides a discharge current path for the energy stored in the magnetic field, which allows that energy to dissipate, rather than appearing as a voltage spike.

The coil current curve can be interpreted to get the opening and closing times of the valve. **Fig. 5.6** shows the voltage difference as a measure for the coil current as a response to a step input on the transistor base. The left and right sides of the figure show the shunt voltage curves of the fail-open and the fail-closed valve. The discontinuity in each curve indicates the armature hitting into its end position. This is the point where the self-induction of the moving armature ends abruptly causing a voltage drop. The pulling of the solenoid happens faster for both valves than the spring pushing the

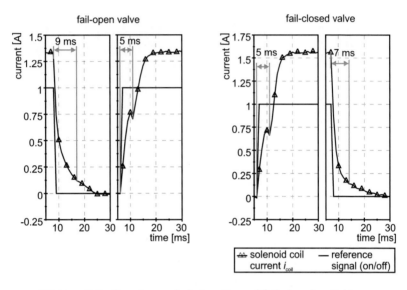

Figure 5.6: Opening and closing time of fail-closed switching valve

armature back. This means that the fail-open valve opens slower than the fail-closed valve. One reason for this is that the spring force pushing the armature back is of course smaller than the pulling force of the solenoid. In addition to that, when the supply voltage is cut off from the solenoid, it takes some time to dissipate the inductive load through the flyback diode and the shunt. To accelerate the dissipation of the inductive load, the electric circuit as presented in section 5.4.3 is enhanced. In this circuit the inductive load is connected to a capacitor where its energy is quickly stored.

5.3 Minimization of pressure steps

The discontinuous flow obtained by switching valves inevitably leads to pressure steps Δp_{step}.

$$\dot{p} = \frac{1}{C_{\text{brake}}} Q \qquad (5.2)$$

$$\Leftrightarrow \Delta p_{\text{step}} = \frac{1}{C_{\text{brake}}} \int_{t_1}^{t_2} Q \, dt \qquad (5.3)$$

The pressure step size is determined by the brake capacity C_{brake}, the pressure difference and valve size, which results in its flow, and the opening time of the valve Δt. The inertia of the flow is also influencing the pressure step. As already stated in section 3.1 it can be neglected for this case. Thus pressure steps are minimized by enlarging capacity or reducing the pressure difference, nominal flow or the opening time. Any of these actions have limitations or drawbacks. Enlarging the capacity leads to lower efficiency, because more supporting cylinder volume is needed per braking. The valve pressure difference depends on the actual supporting pressure and is a characteristic of the self-energization. Reducing it requires the guarantee of a higher minimum friction coefficient. Reducing the nominal flow of the switching valves leads to reduced brake dynamics, which especially effects the initial braking process. The opening time can be reduced to a certain level. Very short impulse periods achieve a partial opening of the valves. The smallest possible impulse length at which the valve starts to open at all can be used to generate very small flow. It is, however, difficult to know this impulse length, since it depends strongly on the port pressures of the valve. To be certain, that the valve opens at all, a realistic minimum impulse length has to be chosen.

For prototype I, [LSM08] presents the pressure step function according to Eq. 5.3 in dependence on the supporting pressure. It calculates the expected pressure step in the brake actuator for a minimal opening time of the valve dependent on the actual supporting pressure. The pressure step function is shown in **Fig. 5.7** on the right side. The simple model on the left of the figure explains the parameters

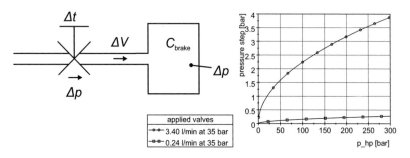

Figure 5.7: Model of brake capacity and pressure step function

used for the pressure step function. The brake capacity C_{brake} not only contains the hydraulic capacity but also takes the mechanical stiffness of caliper and brake pads into account. For comparison: A pressure step of one bar leads to 300 N increase in compression force. Two curves show how a reduced nominal flow of the valve leads to smaller pressure steps. The reduced nominal flow can be achieved by a smaller valve or a throttle placed downstream of the valve. The figure makes it clear that with higher supporting pressures, the control deadband must become larger to prevent instability.

5.3.1 Throttle and bypass

High dynamics at low brake forces, combined with high precision control at high brake forces, can be achieved with two parallel valves.

Valve configurations for pressure control

One of the valves is applied with an orifice or a throttle downstream to reduce its nominal flow. Only for low brake forces and high control deviation it is synchronized by the valve without throttle. This configuration is depicted in Fig. 6.3. In the hydro-mechanical scheme of the first prototype brake, valve BV3 functions as a bypass for BV1 which is supplied with a throttle downstream. For experiments with different throttle sizes, the throttle was made from a screw and could be replaced by opening the plug opposite of the valve as shown in **Fig. 5.8**. The complete valve block is described in section 6.3.2.

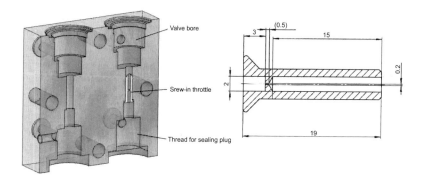

Figure 5.8: Screw-in throttle and valve block

5.3.2 Pulse modulation

As explained in section 5.2.1 pulsing of the valves can be used to generate quasi-continuous flow with switching valves. Pulses are characterised by

- Amplitude

- Width

- Time span between pulses

Each of these properties can be used for pulse modulation. In *pulse amplitude modulation* (PAM) the input signal is encoded in the amplitude of a series of signal pulses with the same impulse length and with a constant frequency. In *pulse width modulation* (PWM) the pulse width corresponds the input signal, while amplitude is maximum and frequency is constant. In *pulse frequency modulation* (PFM) the time between pulses is adjusted to the signal input while amplitude is maximum and width is constant. The pulse modulation techniques are illustrated in **Fig. 5.9**.

PFM and PWM can be implemented quite easily with power transistors, the implementation of PAM is more complex and requires the use of digital/analogue devices. Since the switching valves have no stable position except opened or closed, PAM is neglected here. The modulation for PFM and PWM was implemented by software on the real time control system dSPACE for prototype tests. **Fig. 5.10** shows simplified schemes of implementation. The figure depicts how the reference signal influences the pause between pulses (PFM) and the pulse length (PWM). In addition to that, in PFM the pulse length is set to be dependent on the actual pressure difference at the valve and in PWM the minimum and maximum pulse widths are set according to the actual pressures.

The objective of pulsing the valve is the precise control of small pressure steps and to avoid pressure overshoot. To evaluate the use of PWM and PFM to produce small quasi-continuous flow compared to the nominal flow of the valve, it is useful to look at Fig. 5.2. Pulses of a length below $t_{E,dead}$ do not have any effect. If the length is between $t_{E,dead}$ and $t_{E,dead} + t_{E,stroke}$, a partial opening of the valve is achieved, delivering the smallest volume possible. The minimum

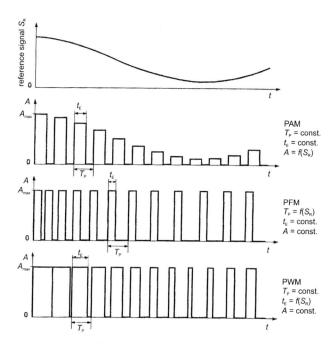

Figure 5.9: Pulse modulation techniques [Wen96]

pressure step is related to this pulse period. If the pulse is longer than t_E, the inductive load leads to a switch-off delay $t_{O,dead} + t_{E,stroke}$. This delay in particular results in the pressure overshoot since although the control deviation is zero, more volume flows into the actuator. **Fig. 5.11** shows the flow Q_V generated by the fail-closed valve over varying pulse length t_E at specific pressure differences. The whole period is 30 ms. The pulse length t_E for partial opening must be in the range of $2 - 2.5$ ms.

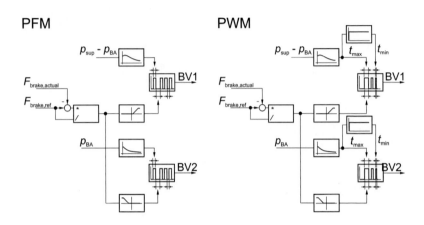

Figure 5.10: Implementation scheme of PFM and PWM

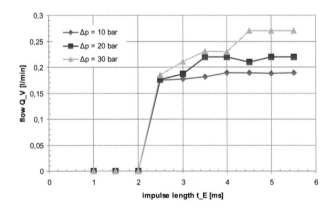

Figure 5.11: Flow of the fail-closed switching valve for varying pulse length and different pressure drops

With pulse-width modulation the pulse width is longer than $t_{E,dead} + t_{E,stroke}$ for most of the time because $t_{E,stroke}$ is small compared to the whole period T. After the shut-off time, according to Fig. 5.2 the valve flow continues for some time $t_{O,dead} + t_{O,stroke}$. Therefore the pressure overshoot is inevitable. Pulse frequency modulation therefore is better suited to avoid overshoot. The pulse width can be chosen such that a partial opening is achieved. Thus the valve flow stops only shortly after the control deviation becomes zero. With this parameterization the valve is never fully opened. While facilitating more precise control the PFM is less dynamic than the PWM because the PWM allows full opening of the valve. Both, PWM and PFM were implemented and tested with the SEHB.

5.4 Proportional 2/2 valves

One way to reduce pressure steps is the use of proportional seat-type valves. Therefore prototype II of SEHB uses proportional seat-type valves for the closed loop brake force control. This section explains the design of the valves used and their static behavior.

Proportional seat-type valves have been developed for special applications where tightness, contamination robustness or cost effectiveness are very important. This is the case for advanced automotive vehicle control systems like ESP (Electronic Stability Program). **Fig. 5.12** depicts a simplified sectional view of a normally open (NO) and normally closed valve (NC). The valves used for this research were kindly provided by Continental Teves. The NC valve is used for the electro hydraulic brake (EHB) and the NO valve has been designed for a traction control system (ASR), where the wheel slip is controlled by braking the wheel in case of loss of traction. In their design, the valves are very similar to the switching valves. They are

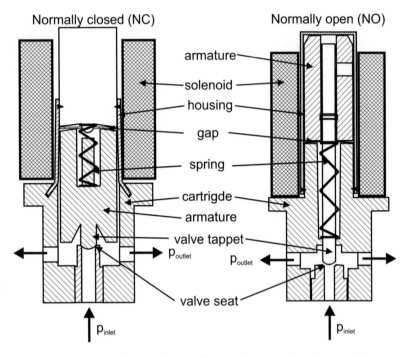

Figure 5.12: Sectional view of normally opened and normally closed valve

also made as cartridges for being press-fitted into an aluminum valve block and have cutting edges which tightly separate inlet and outlet borings in the valve block. The tappet is connected to the armature which is held by a spring in its normal position. At the lower end the tappet has a seat element which closes the connection between valve inlet and outlet. The armature is encapsulated in the housing of the cartridge. The solenoid is a separate part and is mounted by simply putting it over the housing.

Valve configurations for pressure control

The NC valve has a low rate of 40 $\frac{ml}{s}$, the NO valve has 52 $\frac{ml}{s}$ at a pressure drop of 100 bar using DOT 4 brake fluid which has a specific weight of $\rho = 1060 \frac{kg}{m^3}$. Since the SEHB prototype uses HLP46 hydraulic fluid instead, the different specific weight result in a different flow rate, which can be calculated by:

$$Q_{\text{nom}_{\text{HLP}}} = \sqrt{\frac{\Delta p_{\text{nom}_{\text{HLP}}} \rho_{\text{DOT4}}}{\Delta p_{\text{nom}_{\text{DOT4}}} \rho_{\text{HLP}}}} \tag{5.4}$$

HLP hydraulic fluid has a specific weight of $\rho = 870 \frac{kg}{m^3}$. At 35 bar nominal pressure, the NC valve therefore has a nominal flow of 26 $\frac{ml}{s}$ or 1.56 $\frac{l}{min}$. The NO valve has 34 $\frac{ml}{s}$ or 2.04 $\frac{l}{min}$.

The geometry of the gap between armature and core and the spring characteristic is responsible for the proportional behavior of the valve. The term proportional, however, is a bit misleading. It suggests that the valve opening corresponds proportionally to the input signal, which is not the case. The valve opening is not proportional to the solenoid current but is continuous between an opening current where the valves just start to open and an upper boundary current where the valve is fully opened. The knowledge of the solenoid current - valve opening characteristic is vital for the programming of the brake force controller. To achieve partial opening of the valve, the control signal must be mapped on the band between opening current and the current at full opening. The valve dynamics is also an interesting characteristics of the valve because it limits the closed loop control gain of SEHB, as explained in section 3. A dynamic measurement of the NC valve using sweep excitation has been published in [ELSM08]. The bode diagram shows that the valve dynamics is sufficiently high and well damped.

5.4.1 Measurement of tappet movement

Since the tappet is not accessible from outside, to measure the tappet position dependence on the solenoid current without destroying the valve can only be done by optical measurement systems. A laser vibrometer can measure the phase lag between the emitted and reflected laser beam to measure optical distances. Due to the design of the cartridge valves only the NC valve can be opened without destroying it. The NO valve has an inlet filter element which cannot be removed without damaging it. The optical principle allows high dynamic measurement which a flow sensor could not achieve. However, the laser vibrometer needs an optical access to the valve tappet hence the measurement cannot be used for the pressurised valve.

The test bench with the laser vibrometer measuring the tappet position of the NC valve is depicted in **Fig. 5.13**. A measurement system

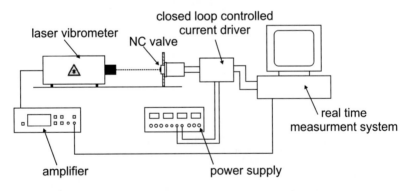

Figure 5.13: Measurement set-up for measurement of current - opening characteristics and tappet dynamics

generates the reference signal and feeds it to the current driver. The current driver supplies a closed loop controlled current to the valve

Valve configurations for pressure control   page 113

solenoid. The laser vibrometer measures the tappet position, which is transformed into a signal between 0 and 10 V by the amplifier and given back to the measurement system.

Fig. 5.14 shows a static measurement of the tappet movement where the solenoid was driven with a slow ramp function. Without pres-

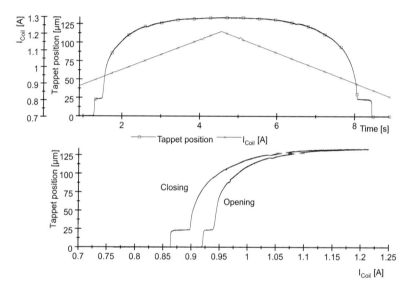

Figure 5.14: NC valve tappet position driven by slow solenoid current ramp function

sure, the valve begins to open at 0.923 A and reaches its maximum opening of 125 μm at about 1.15 A. The upper graph shows the measured position of the valve tappet over time. The lower plot shows the displacement of the tappet as a function of the current for both directions of actuation, opening and closing of the valve. There is a significant hysteresis between the opening and closing of the valve. Due to the friction force, which is always directed against the direction of movement, the current for pulling the tappet out of the seat against the spring is 0.57 A higher than the force for closing it. The opening characteristic is strongly nonlinear. The tappet position gradient is steep for small solenoid current. This makes it difficult to set a small opening of the valve.

The measured step which occurs in the tappet position at 0.92 A and 0.86 A could not be settled conclusively. It is probably caused by friction.

5.4.2 Measurement of valve flow

For parameterization of the brake force control of SEHB, the valve flow characteristics are even more important than the valve opening. It is dependent on the pressure before and after the valve as well as the solenoid current. Changing every parameter individually leads to a four-dimensional map. Using a flow rate sensor between two pressure controlled servo valves allows measurement of the whole map for both, the NC and the NO valve. Comparing the flow at the same pressure drop but different absolute value, the flow is not significantly different for both types of valves. The flow characteristics can therefore be represented by a three-dimensional map, only dependent on the solenoid current and the pressure drop.

The flow rate for both valve types is displayed in **Fig. 5.15** for various pressure drops and solenoid currents [ELSM08]. A characteristic parameter for the valve is the nominal flow at a nominal pressure drop of 35 bar. The NC valve has a nominal flow of 1.61 $\frac{l}{min}$ and the NO valve a flow of 1.95 $\frac{l}{min}$. This data complies with the data from the manufacturer. The flow rate for a fully open valve is shown in Fig. 5.15 according to Eq. 3.3. The hysteresis between the opening and closing path is large considering the small current band in which

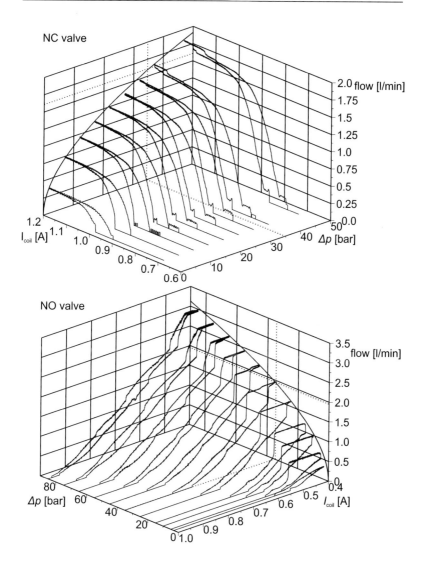

Figure 5.15: Flow maps for NC and NO valve

the valve has continuous behavior. Especially with the NC valve it is difficult to determine the solenoid current for small flows because the flow gradient is steep in that area. The flow gradient of the NO valve is smaller for small openings and the hysteresis is not as large. Therefore the NO valve is probably better suited for control of small flows than the NC valve. This assumption is confirmed by the experimental results in section 6.4.3.

5.4.3 Driver electronics

Each valve is driven by a current controller. The current driver is based on an asymmetric half bridge built up from two transistors (switches) S_1 and S_2 and two diodes D_1 and D_2 [Don05], see **Fig. 5.16**. The solenoid is situated between both transistors and

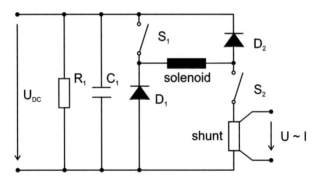

Figure 5.16: Asymmetric half brige of current driver electronics

diodes. Dependent on the switching positions of the transistors, current in the solenoid is built up, free-wheels or is dissipated. If S_1 and S_2 are both closed, supply voltage is applied to the solenoid and current builds up. If S_1 is closed while S_2 is open, the current of the solenoid free-wheels through the shunt and diode D_1. If both switches are closed, the energy of the magnetic field is dissipated in the capacitor C_1 which then is discharged slowly over resistor R_1.

During the build up of current and of free-wheeling, the flowing current can be measured by its voltage drop over the shunt. This signal is used for the current control by a simple switching controller. **Fig. 5.17** shows such a measurement and demonstrates the working principle of the switching controller. The step in the reference sig-

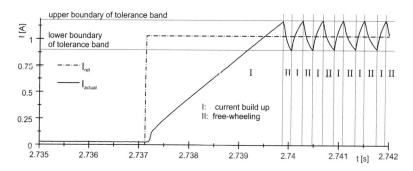

Figure 5.17: Mesurement of current build-up and free-wheeling

nal is shortly after $t = 2.737$ s. Both switches are closed to achieve a current build-up (I) until the actual current reaches the upper boundary of the tolerance band around the reference value. When it reaches the upper boundary, switch S_1 opens and the free-wheeling begins. One can see that during free-wheeling the current drops almost equally fast as it rises in the beginning. During free-wheeling the resistances of the solenoid and the shunt reduce the circulating current by transforming its energy into heat.

In order to cut down the current quickly, the energy stored in the inductance is moved to the capacitor C_1 by opening both switches. The magnetic force applied from the solenoid to the valve tapped is cut off quickly. Since the shunt is not in the current path for this case, no measurement has been taken.

The complete wiring scheme which has kindly been developed and produced by The Institute of Power Electronics and Electrical Devices (ISEA) is shown in **Fig. 5.18** It has two connectors for the real-time control hardware. The reference signal is connected to jumper 10, the actual solenoid current is given back at jumper 9. The solenoid is connected to ports X9-8 and X9-7. For each solenoid one driver electronic is needed.

The marked comparator IC4A is a Schmidt-trigger with an adjustable positive feedback (R87 and R84), with which the hysteresis of the switching controller can be set [TS02]. The inputs of the comparator are the reference signal, which is a signal between 0 and 10 V, and the amplified voltage drop over the measurement shunt. The factor of amplification is $\frac{R71}{R70} = 5.6$ for R71 = R69 and R70 = R68. This means that 10 V represents a solenoid current of 1.785 A. If the amplified voltage is smaller than the reference voltage, the comparator switches positive supply voltage on its output. If it is larger, it switches negative to supply voltage on its output. The output signal is inverted by transistor T12 and applied to the gate of MOSFET T11. T11 is a p-channel-MOSFET which connects if a positive voltage is between its source and its gate. This is the case if T12 is conductive. This means that if voltage at the output of the comparator is positive, MOSFET T11 becomes conductive and the solenoid is connected to supply voltage if T10 is also conductive, which is the case for any positive reference signal. The part between output of the converter and T12 has been neglected so far. This part acts as an AND-element which is only active if both the output of the

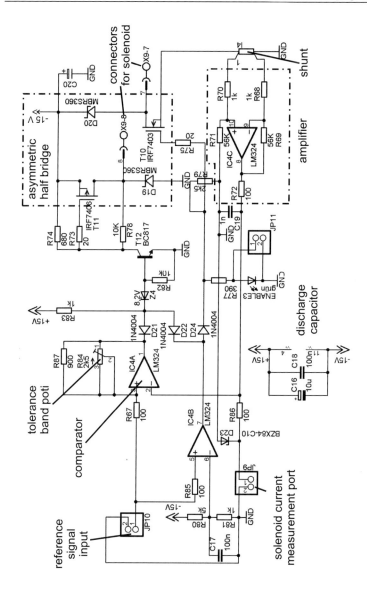

Figure 5.18: Wiring scheme of driver electronics

comparator and the output of the IC4B are positive. The output of IC4B is low when the reference signal is below approximately $-4\ V$. This signal is used for the fast demagnetization of the solenoid. If the reference signal is set to $-10\ V$, both transistors T10 and T11 are switched off. The inductance stored in the solenoid can dissipate against supply voltage in capacitor C16.

Chapter 6

Prototype design and tests

The prototype design of SEHB was done in two phases. In the first stage, it was important to prove the concept of stable brake force control using hydraulic self-energization, since before it was only demonstrated in simulation. Prototype I serves this purpose using as many standard components as possible. It also provides valuable experience for the implementation of an improved customized prototype, referred to as prototype II, which has the full SEHB functionality like active actuator retraction and a fail-open safety concept, which is important for train brakes.

This chapter presents both prototype designs with measurement results in sections 6.3 and 6.4. A major part of each prototype section is the design of the valve block. For the first prototype, conventional automotive anti-lock system fast-switching valves are successfully used. The valves and the electronic switch used as driver for the solenoids are presented in section 5.2.2. One issue with the switching valves are the pressure steps. Another issue is, that the control dynamics are not adjustable to the changing open loop gain during brake pressure rise. New valves used for electro-hydraulic brakes and electronic stabilization programs in automobiles offer some proportional functionality. Therefore the valve block and the driver electronics are completely revised for Prototype II. They were presented in section 5.4.

6.1 Maximum brake force

The maximum brake force is the major performance parameter on which the prototype designs are based. It is derived from the performance specifications of the target application of the research project "EABM" (Einzelrad-Antriebs-Brems-Modul), see section 1. The target vehicle's key data is based on the presumption that a passenger railway car does not necessarily need to be heavier than a comparable road vehicle, for example a bus:

- Regional passenger railways
- Maximum speed: $v_0 = 120 \frac{\text{km}}{\text{h}}$
- Maximum weight: $m = 13.6$ t
- Two pairs of individual wheels, four disc brakes
- Diameter of wheel (new / old): $d_{\text{wheel}} = 920$ mm/840 mm

General performance requirements for railway brakes are standardized in [EN105]. For a maximum stopping distance of 500 m at maximum velocity and an estimated response time of 0.8 s, the brake must provide a maximum deceleration of a little less than $d = 1.2 \frac{m}{s^2}$.

The maximum deceleration force F_d is calculated by multiplying the mass inertia per disc brake times deceleration plus a constant force resulting from slope of $s = 4\ \%$ and gravity g, **Fig. 6.1**. The rotary

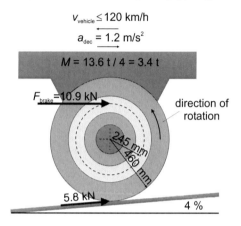

Figure 6.1: Requirement specifications for design of SEHB

inertia of wheels and drives is included with a factor $k_r = 1.1$ in the translatory inertia.

$$F_d = \frac{m}{4}(k_r d + sg) = 5822\ \text{N} \qquad (6.1)$$

The maximum friction force F_{brake} acting on a friction radius of $r_\text{f} = 245$ mm then yields:

$$F_{\text{brake,max}} = F_d \frac{d_{\text{wheel}_\text{new}}}{2 \cdot r_\text{f}} = 10931\text{N} \qquad (6.2)$$

This maximum brake force is the basis for calculations for the prototype design. It is clear that using an automotive brake disk with a brake radius of 115 mm instead of 245 mm, the deceleration momentum produced by this brake is smaller. However, for research on the self-energizing electro hydraulic brake, the brake momentum is secondary. The priority is to reproduce the brake forces and pressure levels in the brake as expected for the target application.

6.2 Brake test stand

The brake test stand consists of a secondary controlled hydraulic motor unit, a fly-wheel connected via safety clutches and elastic couplings and a brake disk on a brake shaft, **Fig. 6.2**.

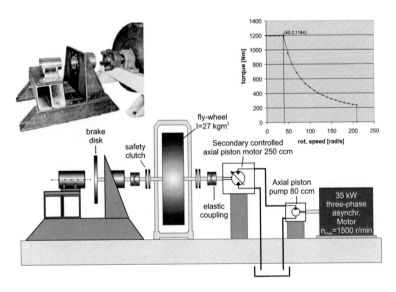

Figure 6.2: Key data of test stand facility

The hydraulic drive of the test stand consists of a Denison fixed displacement axial piston pump OPV4/080 R3Z with 80 cm^3 displacement volume, driven by an three phase asynchronous motor. The Bosch Rexroth A4VSO 250 DS1 variable displacement motor has a maximum displacement volume of 250 cm^3. Motor and pump are connected in an open circuit. The fly-wheel has a mass inertia of 27 kgm^2. It is not equivalent to the mass inertia of the rail vehicle but enough for simulation of small stop brakings. The graph in the upper left of Fig. 6.2 shows the speed-dependency of the maximum load which can be achieved by the drive. The maximum load is limited by pressure relief valves, set to 300 bar. The maximum speed is limited by the motor control unit to 2000 $\frac{rev}{min}$. The picture in the upper right shows the test stand in assembly.

6.3 Prototype I, plunger actor

For the first simplified prototype the SEHB-scheme is modified in such a way that a standard pin-slide brake caliper of a Mercedes Benz W124 E-class can be used.

6.3.1 Hydraulic-mechanic design of Prototype I

In contrast to the previous discussed schemes, the brake actuator is a plunger cylinder which cannot be actively retracted and the pre-stressed spring to initiate the process is not installed. It has a piston diameter of 54 mm. **Fig. 6.3** shows the structure of the test configuration including valves, sensors and control. **Fig. 6.4** shows the brake actuator and the supporting cylinder in detail with some dimensions. The brake radius is 110 mm. Caliper and supporting piston rod are connected to a pivoted bar which conducts the braking forces. The distance from the center of rotation to the supporting cylinder axis is adjustable around 110 $mm \pm 10$ mm.

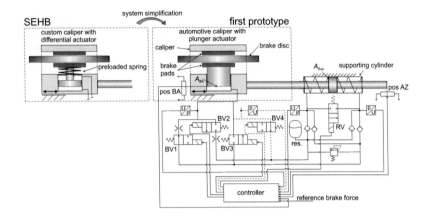

Figure 6.3: I

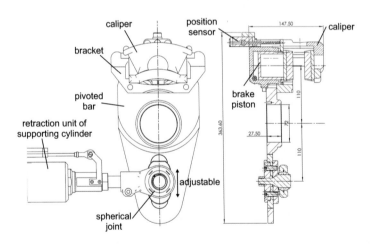

Figure 6.4: Main dimensions of brake actuator of Prototype I

The supporting cylinder has a piston/piston rod diameter of 40/25 mm. As shown in Fig. 6.3 it is centered by springs. However, the retraction unit is realized by one spring only, which is compressed by sliders in both directions of actuation. **Fig. 6.5** presents a sectional view of the supporting cylinder in its middle position. The retraction unit is

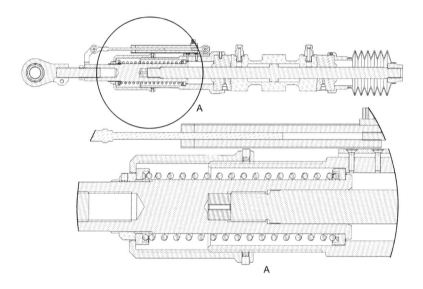

Figure 6.5: Sectional view of supporting cylinder

enlarged in a detailed view. The ends of the retraction spring stand on two slider rings. With the piston pushed out, the right slider is pulled along by the piston rod extension while the left slider is held back by the hollow part connected to the cylinder housing. The spring compression reverses the process when both cylinder chambers are hydraulically connected. If the piston is pulled in, the left slider is pulled along by the piston rod extension while the right part is supported by the hollow part connected to the cylinder housing. Thus, the same spring can be used to bring the cylinder back into its middle position. To overcome friction forces in the cylinder, the spring is pre-stressed in its middle position. This limits the use of the brake for small brake forces, since the supporting pressure starts to rise only when the supporting force is larger than the pre-stressing of the spring.

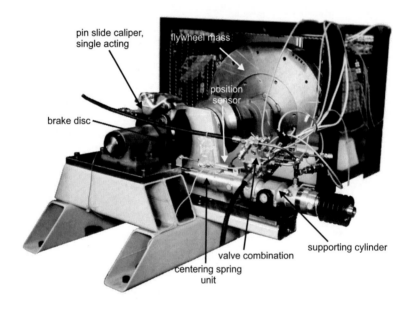

Figure 6.6: Experimental setup of first SEHB prototype

A manually set displacement of the supporting cylinder against its centering spring is used to supply the initial starting pressure in place of the preset spring of the actuator of prototype II. The off-the-shelf supporting cylinder was chosen to be arranged opposite the caliper, as shown in the picture of the experimental setup, **Fig. 6.6**. The retraction unit is attached concentric to the piston rod. Above the retraction unit is the position sensor. The supporting cylinder is connected by a swivel bearing to the test stand's steel frame. Its pressure ports are connected to the valve block incorporating the hydraulic rectifier, a safety pressure relief valve and the control valve combination. From the valve block a flexible hose leads to the brake actuator and a pipe is connected to an open reservoir for compensation of the differential volumes due to the plunger cylinder and oil compression.

6.3.2 Valve block of prototype I

Prototype I has a plunger cylinder as a brake actuator. Therefore it basically only needs two valves, one for the inlet and one for the outlet. The design objective for the valve block was to be able to also use the same valve block for prototype II, which would have a differential cylinder as brake actuator. The fail-closed safety strategy of the brake should be implemented by a fail-open inlet valve of the brake actuator and a fail open outlet valve for piston ring side of the differential brake actuator. It should be possible to exchange the fail-open valves by fail-closed valves. Also threads for exchangeable screw-in throttles have to be placed somewhere in series with the control valves. In addition to the control valves, the hydraulic rectifier, a pressure relief valve and screw-in threads for pressure sensors, accumulators and deaeration screws have to be integrated.

To fulfill these requirements, the valve block for prototype I has a modular design that offers flexibility to gain experience with different valve configurations. **Fig. 6.7** shows the hydraulic scheme of the valve block. The modules $M2$ with fail-open or $M2a$ with fail-closed valves can be exchanged. Modules $M2$, $M2a$ and $M3$ all have the

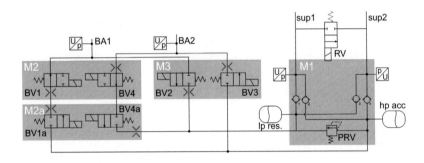

Figure 6.7: Modular design concept of prototype I valve block

same connection ports: high and low pressure as well as left and right chamber of brake actuator. By feeding these vertically through the module and applying the valves horizontally to connect the appropriate ports, a very compact but yet flexible valve block design evolved, where all modules are mounted onto each other. This design offers full flexibility to be able to do intensive experimental studies with different valve configurations. **Fig. 6.8** presents the realized valve block design.

The base module M1 holds all external connections to the supporting cylinder and brake actuator. The hydraulic rectifyier is realized by small Hydac screw-in check-valves *RVE-1/8*, with an opening pressure of 0.5 *bar*. Low and high pressure lines are connected by a *DB3E*, adjustable Hydac pressure relief valve, which is set to 180 *bar*. All screw-in threads are G 1/4 to enable easy exchange of connectors. Fig. 5.8 contains a sectional view of module *M2* and an enlargement of the optional screw-in throttle which is inserted between sealing plug and valve. The actual valve configuration used for prototype tests presented in the next section is depicted in Fig. 6.6. BV1 and BV2 are optionally applied with screw-in throttles.

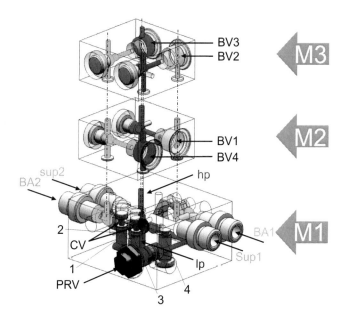

Figure 6.8: Realized design of valve block for prototype I

6.3.3 Test results with prototype I

The following figures present the test results of the first prototype. For each of the control modes, switching control, PFM and PWM, two measurements are shown in one figure. The first measurement, plots a) – d), is the system's response to a reference signal ramp function. The second one, plots e) – h), demonstrates the response to a reference signal step function. The displayed measurement variables and scales of all figures are equal, so that the results can be compared. The displayed measurement variables are, from top to bottom:

- brake force reference signal
- achieved actual brake force
- valve actuation signal BV1, fail open
- valve actuation signal BV2, fail closed
- supporting pressure
- actuator pressure
- position of supporting cylinder

Switching control

The switching controller is implemented in the real time control system *dSPACE* according to the scheme depicted in Fig. 5.3. Upper and lower deadband boundaries were set to 200 N. The throttle diameter upstream of BV1 is 0.2 mm and upstream of BV2 0.3 mm. BV3 is not applied with a throttle and serves as bypass valve to enhance dynamic behavior below brake forces of 1000 N, as proposed in section 5.3.1. The measurement results using the switching controller are displayed in **Fig. 6.9**.

The valve reference signal curves in the Fig. 6.9 b) and f) show the switching controller behavior. The valve is opened as long as there is control deviation above 200 N, and closes as soon as the deviation is below 200 N. This results in very few switching cycles compared to the other methods, but it is important to tune the pressure build-up dynamics by inserting the right throttle size. At the beginning of the step response the actual brake force curve shows some depression. The actual brake force is calculated from supporting cylinder pressures, taking an estimate of the spring forces of the retraction unit into account. Looking at the pressure build-up of the brake

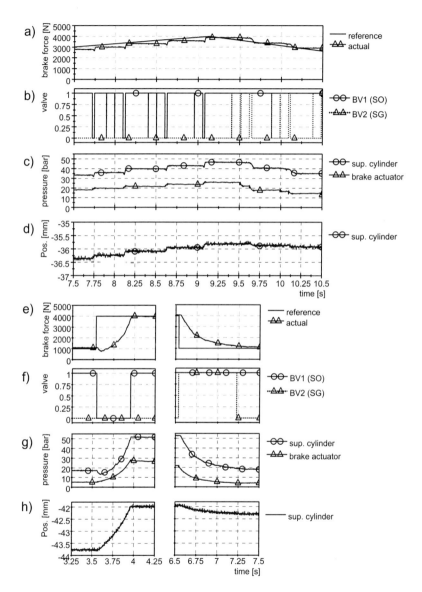

Figure 6.9: Measurement results prototype I (switching control)

actuator, one can see it is instantaneous. This indicates that, in contrast to the brake force measurement, the real brake force must be rising too. The depression in the supporting pressure is caused by friction of the supporting mechanism. The friction in the links and bearings, the sliders in the retraction unit and the supporting piston adds to the reaction force of the supporting cylinder. This means that during pressure build-up, the measured supporting pressure indicates a lower brake force than the one which is actually applied. The lower graph shows the movement of the supporting cylinder. The whole brake force step affords less than 1.2 mm of supporting cylinder movement.

Pulse frequency modulation

The PFM is realized according to the scheme depicted in Fig. 5.10. The control by short pulses of length T_E of the valves allows to operate without the use of throttles at BV1 and BV2 in Fig. 6.3. Therefore a bypass valve is also not needed. The parameterization of the control, however, is more complex. Besides the cycle time T_P, which is the control variable, the pulse length T_E is parameterized in relation to pressure difference over the fail-open and fail-closed valves. This is done on the basis of measurements as shown in Fig. 5.11. T_E is around 10 ms for BV1 (normally open valve) and around 3 ms for BV2 (normally closed valve). The minimum cycle time is limited to $T_P = T_E + 25$ ms because the closing times showed to be too long, if the valve opened permanently. The measurement results using the PFM controller are displayed in **Fig. 6.10**.

Comparing the brake force of the step response of switching controller and PFM, graph e), the PFM controller is slower in pressure build up but a little faster in pressure release. Since BV1 is responsible for pressure build-up, the longer pulse time is the reason why pressure

Prototype design and tests

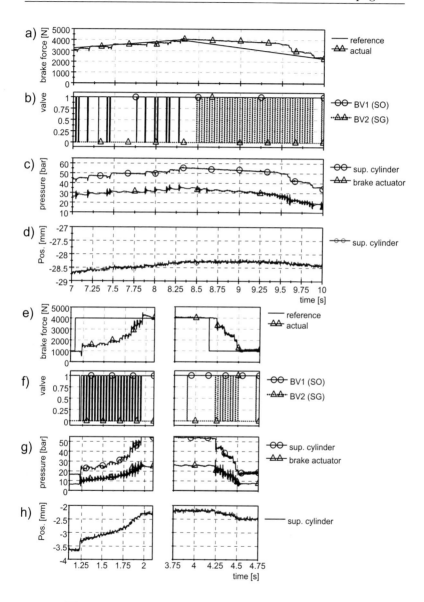

Figure 6.10: Measurement results prototype I (PFM)

build-up is faster than pressure release. The control deviation is smaller in comparison to the switching controller, which is especially evident in the upward ramp, curve a). The PFM needs many more switching cycles, curves b) and f), which may be a problem for the lifetime of the valve. Although the pressure step size is smaller, each switching produces oscillation in the supporting cylinder and especially the brake actuator pressures, curve g). These oscillations also have higher frequency because, without the throttles, the open loop gain is higher.

Pulse width modulation

The PWM control scheme is depicted in Fig. 5.10. Similar to the PFM control, no extra throttles of the valves are needed. The cycle time T_P for every pulse cycle is constant. It is $T_\mathrm{P} = 25\ ms$ for BV1 (normally closed valve) and $T_\mathrm{P} = 20\ ms$ for BV2 (normally closed valve). Precise control of the brake pressure is achieved by setting the width of the control pulse T_E between a maximum and a minimum value. Therefore T_E varies between $3 - 12\ ms$ for BV1 and between $3 - 4\ ms$ for BV2. The measurement results using the PWM controller are displayed in **Fig. 6.11**.

By variation of the pulse width, with the given parameters, a little larger volume flow can be realized than with the PFM, where each pulse is so short that the valve does not fully open. Therefore the step response pressure build up, curve e), is almost as fast as with the switching control. The pressure release time is similar to the PFM control. Interestingly, brake pressure overshoot is avoided more effectively than with PFM. The pulse width modulation is also more effectively in following the ramp function than switching control and PFM. Concluding, one can say that using PFM and PWM, switching valves without throttles can be used well for wide operating condi-

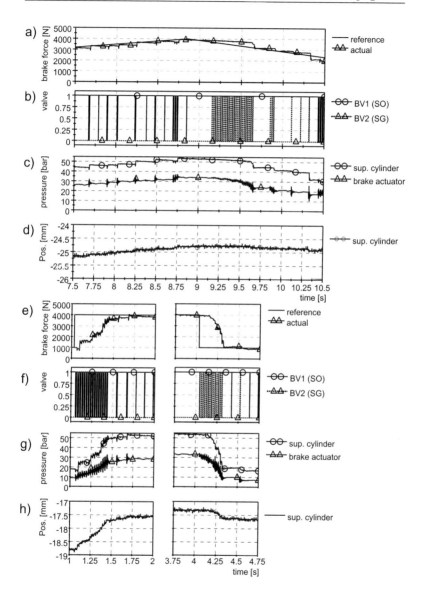

Figure 6.11: Measurement results prototype I (PWM)

tions. The effort to apply the control algorithm, however, is easiest for the switching control. What has not yet been studied so far is how temperature and wear during the valve lifetime influences the switching times and how these influences could be compensated for by the control.

6.4 Prototype II, differential actor

The most important difference between prototype I and prototype II is that there is a differential actuator instead of the plunger actuator. It has been explained in section 4.1 that the main reason for this is to be able to actively retract the brake pads from the disk. A side effect is the faster pressure build-up because of two active pressure chambers instead of only one. The fail-closed safety strategy of the brake is realized by the fail-safe positions of the valves and by the initialization spring which is part of the brake actuator. **Fig. 6.12** gives an overview of the assembly and main dimensions of the caliper of prototype II.

The brake consists of two main assembly groups, the supporting frame and the caliper. The differential brake actuator has a piston diameter of 80 mm a piston rod diameter of 50 mm and a maximum stroke of 40 mm. The supporting frame laterally guides the brake pads and conducts the brake force via supporting brackets into the supporting cylinder. The supporting cylinder unit is reused from prototype I and has a piston/piston rod diameter of 40 mm / 25 mm and a maximum stroke of 100 mm. The brake actuator is attached to the caliper with screws and applies the compression onto the brake pads. To be centric over the axially fixed brake disk, with proceeding wear of the brake pads, the whole caliper must move perpendicularly toward the brake disk. Therefore, in this direction the caliper is guided

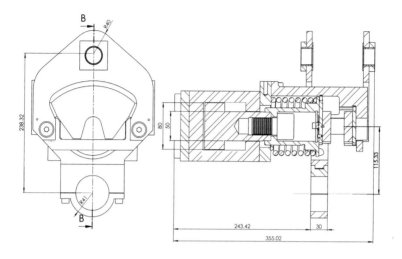

Figure 6.12: Main dimensions of caliper with differential actuator

on pin slides which are fixed in the supporting frame. The brake pads are laterally guided in the supporting frame which conducts the brake force directly into the supporting cylinder.

Fig. 6.13 shows the experimental setup of prototype II. One can see that, in contrast to prototype I, the valve block is attached directly to the back of the block-type differential actuator which provides a very stiff hydraulic connection between valves and actuator chambers. Two hoses connect the valve block with the supporting cylinder. They make only a small movement due to the tilting of the supporting cylinder in relation to the actuator.

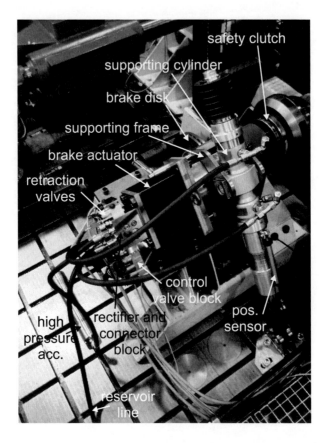

Figure 6.13: Experimental setup of second SEHB prototype

6.4.1 Hydraulic-mechanic design of Prototype II

The hydraulic-mechanic design of prototype II including the setup of sensors and filters is depicted in **Fig. 6.14**. Valves BV1 and BV4 are

Prototype design and tests

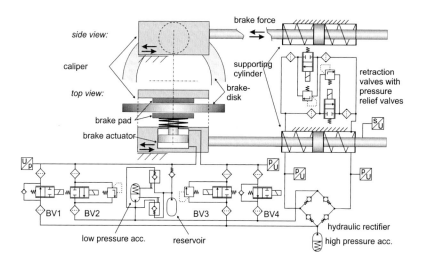

Figure 6.14: Hydraulic-mechanic scheme of Prototype II

normally open (NO) and BV2 and BV3 valves normally closed (NC) to realize fail safe braking in case of complete loss of power. The valves BV1-BV4 are actually doubled to achieve higher maximum flow. It results in higher dynamics at low brake forces. Two opposite faced normally closed valves are connected to the chambers of the supporting cylinder. They are used to retract the supporting piston when no brake force is active. The pressure relief valve symbols which are depicted alongside the NC valves symbolize that at a certain pressure around 200 *bar* the seat of the NC valves is opened against its closing spring. Therefore no additional pressure relief valve is needed in the system.

6.4.2 Valve block of Prototype II

The valve block of prototype II has been redesigned due to experience gained with the valve block of prototype I. The deaeration of the brake system frequently required opening screw-in plugs of the valve block because it was the highest point of the system. For the sake of the press-fitted cartridge valves the valve block is made of aluminum. Consequently the threads wore out. The new valve block design therefore avoids threads in the aluminum parts.

Due to its modular design, the first valve block also required more space than necessary. The valves were attached to different sides of the valve block with the result that many different cables were branching around it. The valve block of prototype II thus has all valves oriented in the same direction. The wires of all solenoids are combined into a multi-wire cable. **Fig. 6.15** shows the valve block for the brake valves BV1-BV4.

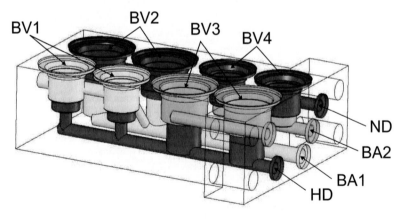

Figure 6.15: Valve block design of prototype II

6.4.3 Test results with Prototype II

This section shows the first measurement results of prototype II. The flow maps depicted in Fig. 5.15 were implemented in the control software. For simplicity, only the piston face side of the brake actuator is controlled using only the brake valves BV1a and BV2a. This configuration is very close to prototype I, which makes it clear to see the benefit of using proportional valves compared to switching valves. Furthermore, it has not been tested if all brake valves have exactly the same characteristics. Using all eight valves with the same parameter setting at an early stage of brake testing makes it more difficult to interpret the control result.

Mathematical analysis in chapter 3 demonstrated that a proportional controller works well for SEHB. But the hysteresis shown in the flow maps poses a problem because it is not possible to define a fixed current value which corresponds to a fixed partial valve opening. The current for the NC valve is especially difficult to define because the flow gradient is steep for small openings.

To handle this hysteresis, the first tests are done using a switching integral controller in addition to the proportional control. The proportional control is working well for large but not for small control deviation due to the hysteresis of the opening current. If the actual brake force does not reach the reference value because the valve does not open due to its hysteresis, the integral controller winds up until the opening current is reached. The switching integral controller only winds up when the actual brake force is in a defined region around the reference value. This limits the integral controller value because it cannot wind up during phases of large control deviation.

Fig. 6.16 shows measurement results of prototype II. The displayed values are

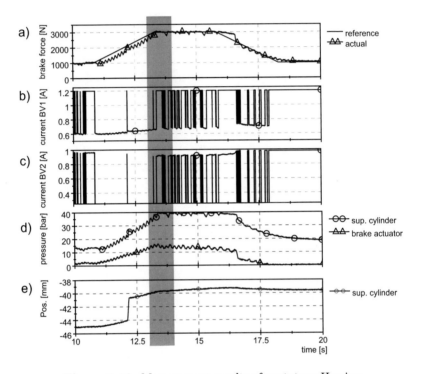

Figure 6.16: Measurement results of prototype II using proportional and switched integral control

a Reference and actual brake force

b Reference value for current of brake valve BV1

c Reference value for current of brake valve BV1

d Pressures in supporting cylinder and piston face chamber of brake actuator e position of supporting cylinder

The reference signal is a ramp rising from 1000 N to 3000 N brake force. The plots show a measurement period of ten seconds.

The measured brake force follows the reference signal with a slight lag of 0.42 s in the upward ramp. For the downward ramp, the actual value is 0.1 s ahead of the reference value. This indicates that the flow gain of the normally closed valve BV2a, which is active during the downward ramp, is larger than the flow gain of the normally open valve. The flow maps in Fig. 5.15 support this explanation. With the NC valves it is more difficult to set a small flow gain because for small openings the gradient of the current-flow curve and the hysteresis are high.

The pressure signals in plot a) and d) of Fig. 6.16 show a distinct ripple. By closing all valves manually, the ripple is still present. The explanation is that the ripple is caused by brake judder.

In the upward ramp at $t = 12.16\ s$ the brake force in plot a) and the supporting pressure in plot d) have a peak. This peak is caused by the high pressure accumulator which hits in its end position. Plot e) shows the supporting cylinder movement which is relatively fast just before $t = 12.16\ s$. During that phase the high pressure accumulator is charged.

At $t = 16\ s$, when the ramp begins to fall, a significant control deviation occurs. During this time the integral part of the controller increases. This can be seen by the fact that for the following control actions of BV2a, although control deviation is relatively small, the current value is higher than just after $t = 16\ s$.

The supporting cylinder stroke is shown in plot e). For the shown ramp rising from 1000 N to 3000 N only 5.5 mm, this is 11 %, of the maximum stroke of 50 mm are needed. Included in these 5.5 mm

are about 3 mm for charging the high pressure accumulator which is only needed once during a brake operation. This demonstrates the high pressure fluid efficiency of the brake.

Fig. 6.17 shows the section of Fig. 6.16 marked by the shaded box. The working principle of the control can be seen well from plot b)

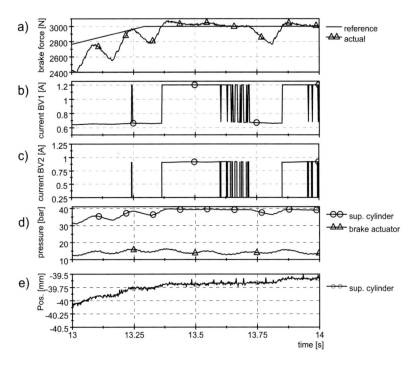

Figure 6.17: Detail of measurement results of prototype II

and c). The tolerance band in which the control output is zero is ± 10 N. If the measured brake force is above this tolerance band, BV2 is switched with the current offset read from the valve flow measurements plus the integral controller value. If the brake force is below the tolerance band, BV1 is switched with its opening current accordingly. The NO valve is closed by applying 1.2 A, while the VC valve is just switched off.

The control behavior using proportional valves is significantly enhanced compared to the switching valves although the results are not easily compared due to the brake judder which occurs in the measurement of prototype II. After intense testing, brake judder occurred in tests with prototype I too. Future research will address this issue in more detail. Normal brakes do not give feedback of the actual braking torque. SEHB uses the actual braking torque as a control variable. The feedback of the brake judder on the normal force probably influences the wear process of the brake disk. However, it has not yet been studied whether brake judder is stimulated or not. One goal of future research is to influence the feedback in order to achieve a better wear process.

Chapter 7

Conclusion

Research on new friction brake principles such as SEHB contributes to the development of innovative brakes with increased functionality adding value to the customer while minimizing energy consumption and maintaining a high safety level. The SEHB is the first closed loop controlled brake which is powered exclusively by self-energization. It is also the only brake which gives feedback about the actual braking torque.

The hydraulic-mechanical structure of SEHB has been developed for a train application with special attention paid to a fail-safe concept, precise brake control, high dynamics, high efficiency and few electrical and mechanical interfaces. **Fig. 7.1** highlights the milestones of the research work on SEHB which is the basis for this thesis. Two strands have contributed to the test results which were finally obtained with the SEHB prototype II: The theoretical analysis gave insight into the dynamics of self-energization and was the basis for controller development. The practical implementation with

Conclusion

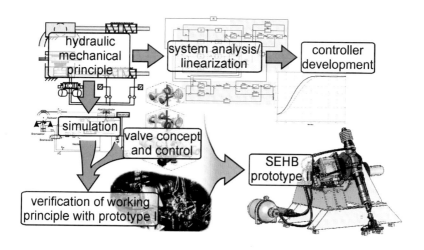

Figure 7.1: Development of SEHB

nonlinear system simulation and a first prototype provided hands-on experience about the brake behavior which inspired the whole development process and helped to identify practical difficulties at an early design stage. The test results verify the working principle of SEHB and point out its high potential for use in trains. The high dynamics offer a means of improving anti-skid braking considerably.

Future research should mainly focus on aspects of system analysis, valve control and mechanical design. A deeper understanding of the dynamic processes and an improved valve control is the basis for enhancing the brake dynamics. The brake judder observed in the brake force measurements is a common phenomenon of brakes. Its causes and effects have been the subject of research in the past. A unique property of SEHB is that brake force ripples are fed back upon compression. Therefore it should be examined how the development of brake judder can be influenced by this. The goal is to find measures to reduce brake judder or at least delay its development and therefore prolong the service life of the brake disks. The most robust solution is achieved by compensating characteristics of the mechanical design which act passively. Active methods can supplement to a more homogeneous wear process. Therefore further theoretical and experimental study of the brake dynamics is a focus of future work.

Another focus is on aspects of mechanical design, particularly on integration and miniaturization. One benefit of SEHB is the reduced weight compared to pneumatic systems including their power supply, reservoirs and distribution system. Considering only the brake actuator installed in the bogie, the research prototypes of this thesis do not yet demonstrate the potential weight reduction which can be achieved by an integrated design. Since hydraulic actuators have a higher force to weight ratio, SEHB can be designed lighter than conventional pneumatic actuators.

The principle of SEHB is not limited for use in trains only. The principle of hydraulic self-energization can also be transferred to other systems with special demands on efficiency and torque control such as road vehicles [LEE+08] or stationary systems. An advantage of SEHB is that it is easily scalable to higher brake forces. To a large extend, valve control is independent of the brake pressure level to be controlled. Especially for heavy trucks, SEHB has potential since compression forces are very high compared to cars. The hydraulic-mechanical structure of SEHB has to be adapted and further developed or simplified to meet the special needs of each application.

Chapter 8

Bibliography

[Abe06] Dirk Abel. *Mess- und Regelungstechnik [Automatic control engineering]*. Aachener Forschungsgesellschaft Regelungstechnik e.V., 2006.

[BB04] Bert Breuer and Karlheinz H. Bill. *Bremsenhandbuch – Grundlagen, Komponenten, Systeme, Fahrdynamik [Brake compendium – fundamentals, components, systems, vehicle dynamics]*. Vieweg, 2004.

[Bec04] Uwe Becker. *Steuerungs- und Regelungsverfahren für fluidtechnische Anwendungen mit Schaltventilen [Control methods for fluid systems using switching valves]*. PhD thesis, Technical University Braunschweig, Germany, 2004.

[Don05] R. W. De Doncker. *Elektrische Antriebe - Skript zur Vorlesung [Electric drives]*. 2005.

[ELSM08] Julian Ewald, Matthias Liermann, Christian Stammen, and Hubertus Murrenhoff. Application of proportional seat valves to a self-energizing electro-hydraulic brake. In *Proceedings of the Symposium on Power Transmission & Motion Control, PTMC'08*, Bath, England, 2008.

[EN105] DIN EN 13452-1 Railway applications – Braking – Mass transit brake systems - Part 1: Performance requirements, 2005. Comité Européen de Normalisation (CEN).

[Fei87] Hans Jörg Feigel. Nichtlineare Effekte am servoventilgesteuerten Differentialzylinder [Nonlinear effects of a servo controlled differential cylinder]. *Ölhydraulik und Pneumatik, O+P*, 31(1):42–48, 1987.

[Fis99] Markus Fischer. *Entwicklung miniaturhydraulischer Komponenten und Systeme am Beispiel autarker Werkstück-Spannvorrichtungen [Development of miniature hydraulic systems]*. PhD thesis, RWTH Aachen Universtity, Germany, 1999.

[Föl94] Otto Föllinger. *Regelungstechnik: Einführung in die Methoden und ihre Anwendungen [Automatic control engineering: Introduction into methods and applications]*. Hüthig, Heidelberg, 1994.

[FRBW+06] J. Fox, R. Roberts, C. Baier-Welt, L. M. Ho, L. Lacraru, and B. Gombert. Modeling and Control of a Single Motor Electronic Wedge Brake. In *SAE World Congress*, 2007-01-0866, 2006.

[GG06]	Bernd Gombert and Philipp Gutenberg. Die elektronische Keilbremse - Meilensteine auf dem Weg zum elektrischen Radantrieb [The electronic wedge brake - milestones on the road to the electric wheel drive]. *ATZ*, 108(11):904–913, 2006.

[Gra99]	Dietmar Gralla. *Eisenbahnbremstechnik [Railway brake systems]*. Werner Verlag, Düsseldorf, 1999.

[HHLS08]	Marcel Hermanns, Martin Hennen, Matthias Liermann, and Thorsten Stützle. Intelligentes, Integriertes Einzelrad-Antriebs-Brems-Modul (EABM) [Intelligent, integrated, single wheel traction and braking module (EABM)]. *ETR - Eisenbahntechnische Rundschau*, 2008.

[HM72]	H. Hesse and H. Möller. Pulsdauermodulierte Steuerung von Magnetventilen [Pulse width modulated control of solenoid valves]. *Ölhydraulik und Pneumatik, O+P*, 11, 1972.

[HSPG01]	Henry Hartmann, Martin Schautt, Antonio Pascucci, and Bernd Gombert. eBrake©- the mechatronic wedge brake. In *SAE World Congress*, 2002-01-2582, 2001.

[Kip95]	Johann Carsten Kipp. Elektrohydraulische Bremssysteme für schienengebundene Nahverkehrsfahrzeuge [Electrohydraulic brake systems for track-bound rapid transit vehicles]. *ZEV+DET Glasers Annalen*, 119:518–524, 1995.

[KPC07]	Hanfried Kerle, Reinhard Pittschellis, and Burkhard Corves. *Einführung in die Getriebelehre. Analyse und Synthese ungleichmäig übersetzender Getriebe [Introduction into mechanisms theory]*. Teubner, 2007.

[Kür93] T. Kürten. *Modellbildung, Simulation und digitale Regelung hydraulischer Antriebe mit elektrohydraulischen Schaltventilen [Modeling, simulation and digital control of hydraulic drives using electrohydraulic switching valves]*. PhD thesis, Mercator Universität Duisburg, Germany, 1993.

[LEE+08] Matthias Liermann, Julian Ewald, Jan Elvers, Hubertus Murrenhoff, and Christian Stammen. Die selbstverstärkende hydraulische Bremse - Integrierte Bremsmomentregelung ohne elektrische Energie [The Self-Energizing Hydraulic Brake - Integrated Braking Torque Control without Electrical Energy]. *ATZ*, 110(10), 2008.

[Lit79] L. Litz. Ordnungsreduktion linearer Zustandsraummodelle durch Beibehaltung der dominanten Eigenbewegungen [Order reduction of linear state space systems maintaining the dominant characteristic dynamics]. *Regelungstechnik*, 27:80–86, 1979.

[LS06] Matthias Liermann and Christian Stammen. Selbstverstärkende hydraulische Bremse für Schienenfahrzeuge [Self-energizing hydraulic brake for rail vehicles]. *Ölhydraulik und Pneumatik, O+P*, 50(10):500–506, 2006.

[LS07] Matthias Liermann and Christian Stammen. Development of a self-energizing electro-hydraulic brake (SEHB) for rail vehicles. In *The Tenth Scandinavian International Conference on Fluid Power, SICFP07, Tampere, Finland, May 21-23, 2007*, Tampere, Finland, 2007.

[LSM08] Matthias Liermann, Christian Stammen, and Hubertus Murrenhoff. Development and experimental results of a self-energizing Electro-Hydraulic Brake using switching valves. In *Proceedings of 6th International Fluid Power Conference*, Dresden, Germany, 2008.

[Mur05] Hubertus Murrenhoff. *Grundlagen der Fluidtechnik, Teil 1: Hydraulik [Fundametals of fluid power systems, part 1: hydraulics]*. Shaker, Aachen, 2005.

[Mur08] Hubertus Murrenhoff. *Servohydraulik - geregelte hydraulische Antriebe [Servo-hydraulics - closed loop controlled hydraulic drives]*. Shaker, Aachen, 2008.

[Ort04] William C. Orthwein. *Clutches and brakes – design and selection*. Marcel Dekker, New York, 2004.

[Per00] Bo N. J. Persson. *Sliding friction – physical principles and applications*. Springer, Heidelberg, 2000.

[Pie94] Martin Piechnick. *Analyse hydraulischer Regelkreise unter besonderer Berücksichtigung von Übertragungsfunktions-Nullstellen und Poldominanzen [Analysis of hydraulic feedback control regarding the transfer function zeros and pole dominances]*. PhD thesis, RWTH Aachen University, Germany, 1994.

[Pro86] E. Prochnio. *Ein Konzept zur Pulsmodulierten Regelung hydraulischer Antriebe [A concept for control of hydraulic drives by pulse modulation]*. PhD thesis, Mercator Universität Duisburg, Germany, 1986.

[RGH+04] Richard Roberts, Bernd Gombert, Henry Hartmann, Dittmar Lange, and Martin Schautt. Testing the

	Mechatronic Wedge Brake. In *SAE Paper*, 2004-01-2766, 2004.
[Rob94]	Robert Bosch GmbH. *Bremsanlagen für Kraftfahrzeuge [Automotive brakes systems]*, 1994.
[RSHG03]	Richard Roberts, Martin Schautt, Henry Hartmann, and Bernd Gombert. Modelling and Validation of the Mechatronic Wedge Brake. In *SAE Paper*, 2003-01-3331, 2003.
[Sem99]	Martin Semsch. Neuartige Mechatronische Teilbelagscheibenbremse [Novel electro-mechanical friction brake]. In *XIX. International μ-Symposium, 29./30. Oktober 1999*, pages 53–74, Bad Neuenahr, Germany, 1999.
[SL08]	Christian Stammen and Matthias Liermann. Verfahren und Vorrichtung zum Verzögern einer bewegten Masse [Method and device for deceleration of a moving mass]. German patent application, 2006, 2008.
[SLS06a]	Christian Stammen, Matthias Liermann, and Toni Schiffers. Hydraulische Bremse mit Sicherheitsfunktion [Hydraulic brake]. German patent, number DE 10 2006 044 021 A1, 2006.
[SLS06b]	Christian Stammen, Matthias Liermann, and Toni Schiffers. Parkfunktion für eine Selbstverstärkende Hydraulische Bremse [Parking brake for a self-energizing hydraulic brake]. German patent, number DE 10 2006 E23 457, 2006.
[SS06]	Christian Stammen and Toni Schiffers. Selbstverstärkende Hydraulische Bremse [Self-energizing

hydraulic brake]. German patent, number DE 10 2006 044 022 A1, 2006.

[Ter92] J. Tersteegen. Pulsmodulierte Ansteuerung von Schaltventilen eines elektrohydraulischen Positionierantriebes [Pulse modulation for electro-hydraulic position control using switching valves]. Technical report, Deutsche Forschungs- und Versuchsanstalt für Luft- und Raumfahrt, 1992.

[TS02] U. Tietze and C. Schenk. *Halbleiter-Schaltungstechnik [Semiconductor circuit design]*. Springer, Berlin, 2002.

[uic04] UIC-codex 541-3: Disk brakes and their application – General conditions for the approval of brake pads, 2004. International Union of Railways.

[Wen96] Guy Wennmacher. *Untersuchung und Anwendung schnellschaltender elektrohydraulischer Ventile für den Einsatz in Kraftfahrzeugen [Analysis and application of fast-switching electro-hydraulic valves for use in automobiles]*. PhD thesis, RWTH Aachen University, Germany, 1996.

Lebenslauf

Persönliche Daten:

Name: Matthias Liermann
Geburtsdatum: 24.06.1977
Geburtsort: Essen
Familienstand: verheiratet, zwei Kinder
E-mail: matthias.liermann@rwth-aachen.de

Schulausbildung:

08/1983 - 07/1987   Stiftsschule Essen
09/1987 - 06/1996   Gymnasium Essen-Werden

Sozialpraktikum im Ausland:

09/1996 - 03/1997   Technische Mittelschule Myjava, Slovakei, organisiert durch die Liebenzeller Mission

Hochschulausbildung:

10/1997 - 03/2004   Maschinenbaustudium an der RWTH Aachen
Studienschwerpunkt: Konstruktion und Entwicklung

Berufstätigkeit:

10/1998 - 02/1999   Studentische Hilfskraft am Institut für Allgemeine Mechanik der RWTH Aachen
07/2001 - 06/2002   Studentische Hilfskraft am Institut für Kunststoffverarbeitung der RWTH Aachen
04/2004 - 03/2009   Wissenschaftlicher Mitarbeiter am Institut für fluidtechnische Antriebe und Steuerungen der RWTH Aachen
seit 09/2007   Gruppenleiter System- und Steuerungstechnik